THE
WILDFIRE
TWENTY

A FIREFIGHTER MEMOIR

HAROLD R. LARSON

FriesenPress

One Printers Way
Altona, MB R0G 0B0
Canada

www.friesenpress.com

Copyright © 2022 by Harold R. Larson
First Edition — 2022

All rights reserved.

No part of this publication may be reproduced in any form, or by any means, electronic or mechanical, including photocopying, recording, or any information browsing, storage, or retrieval system, without permission in writing from FriesenPress.

The views and opinions expressed in this book are those of the author and do not necessarily reflect the official policy or position of the Alberta Government.

Cover photo taken by James Williams.

ISBN
978-1-03-914237-4 (Hardcover)
978-1-03-914236-7 (Paperback)
978-1-03-914238-1 (eBook)

1. BIOGRAPHY & AUTOBIOGRAPHY, PERSONAL MEMOIRS

Distributed to the trade by The Ingram Book Company

For my brothers.

Matthew Engelman, Walt Tipton, Marv Supernault,
Kevin Hedger, Mat Kavanagh, and James Williams.

You are missed.

TABLE OF CONTENTS

Introduction	1
Part I – The Fire and the Flood	
1. Nordegg	7
2. Ground Ignition's 1	15
3. Ditch Beers	25
4. The Bomber Boys of Manning	33
5. Redwood Meadows	41
6. The Yellow Shirt Marines	47
7. The Bridge	57
8. In Memory of a Brother	65
9. Shite Creek	71
10. The End was just the Beginning	79
Part II – Summer Storms	
11. Winter Intermission	83
12. Spring Tranquility	103
13. The Fire Flap – Part 1	111
14. The Fire Flap – Part 2	121
15. GP 43	135
16. Just One More Hour	145
17. Is this the End my Friend?	155
18. Delta Crew	161
19. IC 43	165
20. All or Nothing	175
21. The Beginning of the End	181
22. Reflections	185
Epilogue	193

INTRODUCTION

Try to remember the last time a community was devastated by wildfire. For most Canadians that would likely be the 2016 wildfire that forced over 88,000 people to flee the city of Fort McMurray. Now try to remember the last community that was destroyed by fire before that. Some might recall the time when one-third of the town of Slave Lake burned to the ground in 2011 – most of you, however, will probably not. Like all major wildfires that have impacted communities in the past, when they are burning, they are at the centre of our attention, but as the last of the flames are extinguished, time moves on and so does the rest of the world.

After each such life altering incident, we look back and ask why it happened, how it could have been prevented, and hopefully, what lessons can be learned and passed on to future emergency responders. When Slave Lake was forced to rebuild in the wake of that fire in late spring 2011, an after-action review to learn from this tragedy was formed by Alberta Wildfire[1] called the Flat Top Complex Wildfire Review. Different recommendations were made

---

1 The government branch that encompasses wildfire changed names several times during my career including titles such as Agriculture and Forestry and Sustainable Resource Development. For simplicity I will refer to that branch as Alberta Wildfire.

in hopes of preventing another town from succumbing to wildfire. Some Flat Top recommendations that were implemented included starting the fire season a month earlier on March 1, as well as hiring 10 additional full-time wildfire rangers throughout the province. Before the Slave Lake event, a hiring freeze had taken hold in wildfire circles and full-time work was difficult to find.

The Flat Top review also identified the need for better trained crews who could fight fires on the scale that could destroy entire communities. At that time, there were essentially two types of wildfire crews in Alberta: initial attack that included Helitack and Rappel crews, and sustained action made up of Firetack and Emergency crews. Most crews were successful in their own ways. However, there was a missing cog in the Alberta firefighting machine that was needed to fight such massive, out-of-control wildfires like the ones that destroyed so much of Slave Lake and Fort McMurray.

Initial attack firefighting crews, Helitack, and the now retired Rappel program were proficient at stopping small fires, but once a fire grew to a certain size or intensity, the crew's rate of success was drastically reduced. It would be unfair to think that one helicopter and a handful of firefighters could stop a raging forest fire once it developed to a certain stage. So, when a wildfire became more than what an initial attack crew could contain, they would usually leave and wait for the next wildfire to start in hopes that they could stop that one before it became overwhelming.

To fight the large uncontrolled wildfires, also known as campaign fires, a team of overhead staff, usually made up of career government employees, would be assembled to take on such functions as operations and logistics. The firefighters, mostly comprised of Firetack or Emergency crews, would tackle the fire as the Incident Commander (IC) saw fit. Most of Alberta's Firetack and Emergency crews consisted of men and women who only got hired when there were wildfires to fight. Firetack and Emergency crews can be great firefighters – but when you need a large, organized workforce it is not easy to find crews that possess a high level of experience *and* have the necessary skills that are only developed from working and training together every day.

So it was decided that a new 20-person "Unit crew" would be introduced into Alberta that would be modelled after the 20-person Unit crews that had been established in British Columbia. These BC Unit crews had been crushing campaign fires for decades and Alberta saw potential in a similar program. The formation of this new crew would consist of five sub-crews working together under the supervision of one leader to fight those fires that were beyond initial attack capabilities

In 2012, the first Unit crew was established in the Grande Prairie District. During that wildfire season, major wildfires raged throughout Alberta, and the Grande Prairie Unit crew was truly tested. Under the leadership of a young firefighter named Ryan Archibald, the first Albertan Unit crew did an outstanding job. Their efforts earned the birth of two additional Unit crews, one in Slave Lake and the other in my home district of Peace River.

The second the news broke that Peace River was creating a Unit Crew, I jumped at the opportunity. I phoned the wildfire ranger in charge of the new program to ask about the new leader position and before that phone call was done, I was hired as the first Peace River Unit crew leader.

My journey with wildland firefighting began when I was 16 and before I made that phone call, I already had 13 years of experience in my field: five years as a contract firefighter in British Columbia then eight more years working for Alberta Wildfire in the Peace River District in Northern Alberta. Five of those years I had led a Helitack Crew while the previous year I worked as a man-up supervisor leading Firetack and Emergency crews on large-scale fires. I was told that I was the first person on their list and with my previous experience and dedication to the job, I was exactly what they were looking for.

It was spring 2013 and I was in my final semester of college en route to earning a Forest Technology Diploma from the Northern Alberta Institute of Technology in Edmonton. I had been hoping for a full-time career where I could put my diploma to good use but after landing the Unit Crew lead, it was too good of a position to give up for any industry or government desk job. So once I finished my final semester and obtained my diploma, I headed north to start my new position with Alberta Wildfire.

I did not exactly know what I was getting myself into but was never one to shy away from a challenge, especially in the field of wildland firefighting that had become my passion. During my previous experience as a man-up supervisor,[2] crews would rotate in and out of my command every 10 to 14 days. Having little control with the training or development of the people I worked with, I was often frustrated at the consistent lack of results and effort that I saw. With this new Unit Crew, however, I could immediately see the potential of having 19 personally trained firefighters working beside me throughout the fire season.

Even with my experience and proven leadership abilities, I knew I was venturing into uncharted territories and would need all the help I could find, especially knowing I would be getting a group of new firefighters to train. After finding out I was about to embark on this new path, I recruited my friend and college

---

2   Man-up supervisor is a single resource used to lead other crews on fires and depending on one's skills can be used in various roles such as a strike team leader or division supervisor. I would highly recommend to anyone who wants to become a well-rounded firefighter to do this role for at least a season.

classmate Andrew Farley. He had worked on the Grande Prairie Unit Crew the year before and I helped him transfer to the Peace River Unit. I did not know then how much of a clutch move it would prove to have my 18-year-old friend on the inaugural Peace River crew.

The year 2013 marked the tail end of a rare period in Canadian wildfire history, much of it due to the high priced black gold that flowed beneath Alberta's soil, making it one of the richest places in the world. With wealth came the potential for unlimited firefighting resources such as labour, heavy machinery and aircraft, all aided by innovative technology. This was not the case for other places in the world or other provinces in Canada. Fortunately for the Peace River Unit, firefighting resource allocation and tactics would not be limited by restrictive budgets. As Joel Pecotich, one of the Australian firefighters who worked on the Unit, would sum it up years later: "It was like firefighting with all cheat codes enabled."

The thoughts, ideas, and opinions that are shared in this story are solely those of the author and in no way reflect those of Alberta Wildfire. (I had to say that.) I have written this story to celebrate the men and women that I worked with and have done so to the best of my recollection. By consulting the hundreds of pages of notes and documents I recorded, I tried to be as accurate, and hopefully entertaining, as possible.

On March 1, 2013, the Peace River Unit Crew was born. That spring I would be joined by five first-time sub-leaders, each with only two years of Helitack experience, Andrew Farley, and 13 rookies. Over the next eight months we would fight one of the nation's most devastating natural disasters, creating an environment that would test our resolve and every ounce of our resilience. The courage needed brought us together as professionals and as friends more than I thought possible. This is the story of the dedicated men and women that stood together against the fear of the unknown. None of us knew it then but the Peace River Unit experience would become an epoch in our lives, something that the Wildfire Twenty can now look back on and be proud of.

1

NORDEGG

"We're not going to make it man."
- Elias Niederkorn, Saskatchewan Park Fire 2018

"Peace Unit Leader this is Ops," came the voice of the operations section chief over the handheld radios that each firefighter was carrying.

"Go for Peace Unit Leader," I answered into my radio. I was working with Delta crew leader Jordan Sykes, one of my five Unit sub-leaders. We were moving several lengths of 100-foot, 1 ½-inch nylon hose up a dozer line[3] to extend the hose lay that we were establishing around the south flank of the fire. To the west

---

3   Dozer lines are fuel breaks created by dozers that blade the earth to mineral soil thereby removing a path for the fire to travel. Dozer groups consisting of two or three bull dozers are led by a dozer boss that flags out the path for the dozers to follow. These paths usually hug the fire's edge and are typically done at night when the fire behaviour is at its lowest, and firefighters are clear of the trees they push over.

of the guard, towering white spruce trees bordered our path. To the east, a charred stand of smaller black spruce stood among the smouldering ground. Both of us were covered in a thick layer of ash. We'd been working in the black since we had started our shift six hours earlier at 5:30 am.

"We need you to assemble your burn team and get to staging," replied the voice over the radio.

"That's all copied," I transmitted. During the morning briefing, the Incident Commander had asked if any of the crews had burning experience in case we needed to light a back burn to contain the fire that was threatening the small community of Nordegg, two kilometres to the northeast. I was the only leader to volunteer their crew that morning for the task. The tactic of back burning, fighting fire with fire to remove unburned vegetation or "fuel" from a strategic point, was commonplace for most wildfire services around the world but had slowly faded from the Alberta tactical repertoire. I had been fortunate; I learned how to effectively back burn from fighting fires in British Columbia and Australia and had also recently conducted dozens of prescribed burns with my crew in the Peace River Valley back in our home district.

It was May 12, 2013, now the second day on the fire for me and the Peace River Unit Crew and seven days since the fire had started. Five days before we arrived at the wildfire called "The Fish Lake Fire," it had blown up on three separate occasions and was now being held at just over 107 hectares by 90 firefighters and various types of helicopters, aircraft, and heavy machinery.

It was warmer than usual for mid-May in these rolling hills east of the Rockies. The temperature was in the mid 20s and the winds blew consistently from the southwest. The humidity was low. And *crossover* (a term used when extreme fire behaviour is assisted when air temperature is higher than the relative humidity) was certainly possible during peak burning times in the day. Not only that but the forest was dry. Really dry. Both the Fine Fuel Moisture Code and the Duff Moisture Code were in the "very high" range of the Fire Weather Index. Rain had been scarce and there had been only low levels of snowmelt that spring. Needless to say, there was more than cause for concern.

This was the first time we had fought a wildfire together since the Unit's inception a few weeks earlier. It was as much a test as it was a learning experience for our crew, and I knew that we would have to crawl before we could walk. 16 of the 20 firefighters were available to make the deployment with the remaining four still attending training camp. I was aware that my 13 years of firefighting experience was greater than that of the other 19 firefighters combined. My five sub-leaders had only two seasons of Helitack under their belts each, so fighting a large fire and leading others was as much a learning experience for them as it was for the rookies who were fighting a forest fire for the first time. Up until this deployment, I hadn't even met the five who joined the fire en route after finishing rookie training camp in the mountain town of Hinton, 175 kilometres north. But we were all in it together – and we called ourselves the Peace River Unit Crew.

For the previous two days, our Unit was sent to extinguish the south flank of the fire. The winds were coming from the southwest, helping to blow the fire away from our containment line. But the winds made for challenging conditions for the firefighters trying to hold the opposite flank. Our section was split into two groups divided by an established gravel road called the Forestry Trunk. On the south edge of the Forestry Trunk was a recently harvested cutblock that the fire had spotted into and grown into a three hectare headache. On the north side, the cutblock transitioned into a black spruce forest then into larger mature white spruce. A dozer guard had been dug from the Forestry Trunk and ran north along the fire's edge giving us a suitable mineral soil guard to work on and anchor from.

I now had to decide who would be on the requested four-person burn team. The issue was: I had nine rookies on the fireline who had yet built the trust needed to leave them in the care of my new sub-leaders whose leadership skills were still being tested.

I radioed Alpha and Bravo leaders James Williams and Marco Alexandrino, and told them to go at staging.

"Jordan, you go as well," I said to my 6'4" friend that stood beside me. Andrew Farley, Warren "Warhammer" Keeler, the crew's only experienced red hat, and the Delta leader were helping on the west flank of the fire and were unavailable. David King (DK), the last of the sub-leaders, was supervising his Charlie crew that was working to contain the fire in a cutblock bordering the Forestry Trunk Road which could not be abandoned. That meant I would have to send a rookie.

"Shit, who else is good," I asked. I had only worked with the crew for a couple of weeks and was still figuring out who was going to be my most reliable firefighters.

"I'll take Dad. I worked with him on the burns and he was great," replied Jordan. He adjusted his bright blue hard hat to better sit atop his thick auburn hair that consisted of hundreds of tightly curled locks. He was referring to Sam Hetherington, a 25-year-old that was the oldest of the rookies, thus the nickname.

"Sounds good," I said, trusting his judgement.

During Jordan's rookie year in 2011, we worked on a seven-person Helitack crew and had become good friends. At 23, he had a degree in Physical Education from the University of Alberta and wanted a job where he could travel around the province where he grew up. Wildland firefighting fit perfectly with his desire to work in the outdoors and travel at the same time.

Jordan's hurried his lanky frame down the dozer guard towards our firefighting trucks leaving me alone with my task. I picked up the eight lengths of hose by their metal ends and continued to drag them farther up the dozer line. Each length of hose was soaked making them heavier than their 15-pound (6.8 kg) cargo weight. The treeline on either side opened into a grassy area, then after another 20 metres, the forest closed in once again around me. Towering mature white spruce trees swayed back and forth in the strong wind, causing their trunks to creak and moan. The winds now blew northeast, helping to keep me and my crew from eating too much of the black ash and smoke that seemed to endlessly find their way into our lungs.

When I came to the end of the hose lay, six of the rookies had run three separate hose lines off the main line and were spraying the ground trying to accomplish our goal – extinguishing the burning ground 15 metres from the fire's edge. The rookies were paired up into three groups, each with a hand tool and a nozzle. While one firefighter used the nozzle to spray the hot areas, their partner would help move the hose through the forest or chop up unwanted roots and down debris that was in their way. The fire had been burning here for almost a week and some of the hot spots had dug in deep, requiring demanding work and precious time to put out.

I continued along the dozer guard dropping the 100-foot nylon hoses, one after another, adding line to the area where the crew could extend the main hose line and work into the smouldering forest. Scouting farther ahead, we would need another couple hundred feet of hose to reach were the forest turned into another recently harvested cutblock. Once we reached that cutblock we would have the south flank of the forested area contained and could then move on to another objective.

When I walked back up the dozer guard to check on my crew, the six firefighters with their shiny red hard hats were working hard, heads down, and digging in the charred ground. All crew member firefighters wore red helmets to distinguish them from the sub-leader (blue) and leader (yellow). This was their first wildfire and it was easy for them to become focused on their tasks and not see the big picture. It was not their fault but just the way of the job until they gained enough experience to delineate the forest from the trees.

Andrew was the elder of the Fulthorpe brothers who were both rookies, held dual-citizenship, and lived part time in California. He moved from one hot spot to another while his partner Michael (Soso) Sorenson helped move the hose through the tangled mess that is the forest. Soso represented the occasional East Coaster who ventured from their fishing communities to find work that seemed so difficult to get on that side of Canada.

Several seconds later, silent as an assassin, one of the 20-metre-tall white spruce trees toppled and crashed right where the two men had just been working, missing them by less than 3 metres. The wind combined with the burned-out root system caused the giant woody vegetation to fall.

"Everybody to the dozer guard!" I yelled to the firefighters who were staring at the fallen tree. All six of their heads popped up in unison at the sound of my voice, reminding me of meerkats looking around for danger. "Now!" All stopped what they were doing and rushed towards me. Another tree farther into the burn crashed to the ground sending sparks rushing into the air. The root mats at the base of the standing trees now moved up and down in rhythm with the swaying of their tops.

The radio buzzed with incoming calls from different leaders and helicopter pilots. The escalating winds had caused the fire on the opposite flank to pick up and threaten the containment line that was being held by the other firefighters that were too far away to be seen by us.

As the six rookies followed me along the bare earth dozer line, I radioed DK and Warhammer with instructions to lead the remaining firefighters back towards my direction. I found a spot close to the shallow pond that we had been using as a water source that was clear of any trees that could fall on us. We knelt while listening to the constant chatter on our radios. DK and his three rookies joined us minutes after they had got the call to regroup while it took the Echo crew a bit longer to get back from the southwest flank where they had been helping.

"Man, a huge tree almost fell on me. It was crazy," Andrew said to his brother Adam who had just arrived with DK. The two brothers had lost both parents to cancer in the last year and now more than ever Andrew felt the need to look after his younger sibling.

Above us in the bluebird sky, a fixed-wing Cessna made a low pass over the billowing smoke on the other side of the fire. The Cessna "Bird Dog" sounded its loud yelping siren indicating that an incoming retardant drop was imminent. Moments later, the much larger air tanker soared directly along the same path that the Bird Dog had just flown, opened its belly, and released thousands of litres of the blood-red mixture of water and ammonium salts. The long mass of retardant hit the treetops with tremendous force, smothering the forest below in the red soup with hopes that it would stop the fire from its impending path. Millions of droplets of water dispersed into the surrounding atmosphere, creating a red mist that slowly settled onto the foliage below.

"Peace Unit Leader this is Ops. Get your crew to a safe spot. The fire's going to take a run," transmitted operations over the radio.

"Copy that. Already done," I transmitted back. I took my faded yellow hard hat off and wiped the sweat from my face as I counted my crew again to make sure they were all there. The weather had turned against us. Fire is driven by wind and with the winds increasing, the fire was about to burst through the containment lines. We would have little choice but to wait until the wind event was over and it would be safe for us to work the line.

2

GROUND IGNITION'S 1

"All I see are ones and zeros."
– Jordan Sykes

In a short time, Jordan's long legs allowed him to cover the distance of the cutblock to the staging area where his firefighting truck was parked. His height was a constant advantage that enabled him to step over downed trees and brush with ease, obstacles which at times slowed the shorter firefighters. Resting his back against the door of his white work truck, he unscrewed the top of the water bottle he had retrieved from his pocket and chugged it, quenching his animal thirst.

As Marco, James, and Dad walked over the burned remnants of the cutblock, their boots stirred up the black ash that covered the ground, causing it to be carried away in the wind. The sight of the three men walking towards him through the post-apocalyptic backdrop caused Jordan's lips to form into a smile. He was never one to take for granted how unique the firefighting experience was and this scene once again reminded him of that fact.

"This is awesome. I can't believe I was chosen to be on the burn team," said Dad as he approached the work trucks.

"Yeah man. This is going to be sweet," replied Jordan. "Let's load up Delta truck and head to main staging for instructions."

Marco climbed into the driver's seat. At 29, he was the oldest on the Peace River Unit. Born in Portugal, Marco immigrated with his family to Canada when he was young, but retained the typical physical traits of his homeland: tanned skin, jet black hair, and thick facial stubble.

Jordan sat shotgun while James and Dad filled in the back seat. The four then took off in the white Chevy 3500 to the main staging area at the gravel pit up the road.

After making the staging area, they parked beside rows of other trucks and ambled into a white ACTO trailer that had been brought in as a command post for the overhead staff. There they were met by the operations chief, whose role was to plan out the strategies that would contain the fire. He had worked on and off for the government for years and was assigned as operations chief to fight the rising fire in this area.

The ops chief was much larger than the four other firefighters and took up most of the space in front of the map that outlined the fire perimeter. He explained two possible burn-off strategies for the burn team he named "Ground Ignition's 1." The strategies were to be implemented if the wind was to shift 180 degrees, causing the fire to jump the current containment lines. His hope was that the burn team would not be utilized, but this wasn't his first fire and he knew that with the winds picking up the odds were against them.

The first possible scenario was if the fire made a run towards Nordegg. The ignition team would back burn off the primary containment line, a 200-metre-long dozer line that was being built north of the head of the fire in a cutblock. This strategy of back burning would remove the fuel from the containment line and, with no fuel to consume, the main fire would hopefully stop at the area that was burned off. This would be the final effort to stop the fire from burning over the Forestry Trunk Road (FTR) and into the continuous large timber that grew between the FTR and the town.

The second scenario was if the FTR was burned over. A secondary dozer line was being constructed to the northwest a kilometre away from the fire's edge. The ignition team would drive ahead of the fire and back burn towards the head using the gravel road as their anchor point.

They would have support from three Firetack crews as well as the Grande Prairie Unit crew that was setting up sprinklers along the primary containment line.

Since Nordegg was less than two kilometres to the northeast of the fire, it was decided that the FTR would be the trigger point for the town's evacuation. If the fire crossed that point, the people that lived there would have to leave. Three days earlier, Alberta Wildfire working with local authorities had ordered the town's 300 people to be on a one-hour notice and that the order to evacuate could come any minute.

When the briefing was complete, the ops chief instructed the ignition team to check out both containment lines so they could be oriented if they got the call. The primary dozer guard was easy enough to see from the FTR and seemed a straightforward area to back burn. The smoke coming from the fire to the south seemed far enough away to not be an immediate concern, so they decided they had time to check out the secondary containment line.

Once they reached the secondary containment line, it was apparent that it was not a viable option to burn off. The dozer guard was surrounded by mature timber and thick undergrowth on mountain side. Jordan was confused as to why they thought it was a good idea to build the dozer guard in that location and was now hopeful that burning off from there would not be an option. It also didn't make sense to him why this guard was so far away from the main fire to begin with. This was his first fire as a sub-leader, and he was not yet willing to question orders from superiors. If he was ordered to burn off from the secondary containment line, he would. Whatever the case, they had now scouted both areas and had to wait for instructions.

Fifteen minutes later they got the call. "Ground Ignition's 1. Get to the primary containment line and keep the fire from crossing the road," radioed the ops chief to the ignition crew that was minutes back from the head of the fire. Marco pressed down hard on the gas causing the tires to spit gravel. Turning onto the FTR the crew could see that the wind-driven fire was racing across the cutblock coming close to the gravel road they were now on.

In the sky, four 802s – small fixed-winged aircraft that could skim on lakes and drop water – crab walked in the strong wind and released their loads one after another in attempts to suppress the flame front. The wind was blowing at 45 kilometres an hour, causing the 802s' retardant loads to disperse in the air, doing little to slow the fire.

The 802s returned several minutes later with full tanks after skimming on a lake a short distance to the east. Jordan calculated that the fire must have been moving at an incredible rate by how much closer the planes flew over him on their next run.

Scanning to the west of the FTR, Jordan watched as a glowing ember floated over the gravel road and landed in a patch of cured grass. "Marco, there," he shouted while pointing in the ember's direction. Marco slammed on the brakes, and Jordan jumped out, running to spot where the ember had created a small fire. Using his size 13 boots, Jordan stomped on the burning grass, smothering it with dirt and putting the flames out. But dozens of embers still drifted over the road creating small fires in the grassy area around him. Dad, who had followed Jordan's lead, had gotten out of the truck but was unable to stop the fires that were spreading around them.

Jordan realized that they were not equipped to stop the fire that had jumped the road and needed to get back to safety fast. "Get to the truck," he yelled to Dad. Both men sprinted to the Chevy that was now almost completely covered in smoke from the main fire that was coming at them. Back burning was no longer an option for them. The fire was cooking towards the FTR and they did not have the resources to stop it.

Not having the time to turn the truck around, Marco threw the truck into reverse and sped up attempting to get out of the smoke that was blacking out the sky above them. Their visibility was reduced to almost zero and there was real cause for concern. Jordan hoped that any other firefighters that were fleeing the scene were driving back to staging and not towards them. With the lack of visibility and the speed in which they were racing, a head-on collision with another firetruck would be fatal.

Driving partially blind for what seemed like minutes, the truck forced its way out of the thick smoke and back onto the south side of the excursion that had crossed the road. They were clear of the smoke and could see the wind-driven fire ripping towards the FTR. Within minutes, the fire followed a path of tall timber through the cutblock. It was gaining momentum. A stand of large trees that lined the road to their west burst into a fully involved crown fire. Quicker than anyone could react, the fire climbed to the tops of the 15-metre white spruce and launched crimson, orange, and red flames far above the tree canopy, sending countless embers well over 100 metres in front of the head of the fire.

Two Firetack crews scrambled to run hose from a water truck – but it was too late. The wind had pushed the fire faster than expected and the entirety of the stand was ablaze. The water coming from the hose did nothing – and the realization of having to let the stand of trees burn was a sobering reality. The oppressive heat from the flame front caused the firefighters to abandon the head of the fire. The fire had now crossed the road and trying to save the timber would be foolish.

With the fire's current spread rate, the option to back burn from the secondary containment line was abandoned. The ground crews weren't equipped to stop this fire. And the air tankers were getting battered by the winds, reducing the effectiveness of their retardant drops. The sprinklers that had been set up ahead of the fire were ineffective as the wind carried embers high above their reach, starting more fires on the far side of the primary containment line.

The only thing left for the ops chief to do was to get everyone off the line and make the call he didn't want to: the call to evacuate the town of Nordegg.

The dispatchers heard the call from operations and began to put the evacuation order into effect. Each resident was either called or notified in person that they had no option but to leave their homes. Families loaded into their packed vehicles and drove north out of town seeking refuge from the wildfire that was burning towards them.

Crews began showing up to staging looking for direction from Ops. Jordan spotted a nodwell, a heavy piece of machine equipped with a water tank, sitting idly on the side of the gravel staging area. "Hey Ops, can we use that?" he asked.

"Take whatever you want," the chief responded.

The four firefighters of Ground Ignition's 1 hurried over to the nodwell and asked the operator to follow them along the flank of the fire that was now east of the FTR. Loaded up with their hand tools they followed the path of the fire, spraying and working the flank, making sure that no fire would get behind them. This nodwell was made by a farmer and housed a wide opening in the top of the water tank. When the tank was empty, they could ask for a pilot to fill it up with their helibucket. This way the four firefighters would not have to stop and wait for water.

Each of them felt adrenaline coursing through their bodies as they followed the flank and did what they could to stop the fire from spreading. The head of the fire, however, moved faster than they could. The flame front disappeared over the ridgeline ahead of them. Helicopters with their buckets soared over them as they chased the head of the fire.

"We are pulling everyone off of the fire," the ops chief radioed to me.

"All copied. I have my crew and we are leaving for staging now," I radioed back. Doing another head count I made sure the rest of my crew was at our trucks before we loaded up and headed for staging. The radio continued non-stop with chatter from aircraft and other crews relaying updates and objectives. The call for evacuation had been made and that meant all crews would need to be accounted for.

I led the two other trucks towards staging. The red haze from the head of the fire glowed in the distance. Minutes later we arrived at staging. No one was there, just a barren sand parking lot. The overhead staff must have all pulled out so we would have to do the same. The radio was jammed full of orders, and messages for other crews made it impossible for me to confirm where everyone was.

Aircraft continued to soar over us and drop fire retardant on the head of the fire in attempts to stop its growth. One of the rookies in my truck opened the door and jumped out of the truck. Letting his excitement get the better of him he took out his cell and videoed the tankers soaring above us.

"Get back in the truck," I yelled to the rookie who stood in front of our truck filming the action. Caught up in the moment – or not hearing me above the roar of the wind or the plane engines above – he continued to film the aerial display.

"Get in the fucking truck now!" I boomed. He seemed startled and snapped back to reality. He ran back into the truck. "Fuck man, don't ever do that again," I said to the clueless rookie in the back seat.

Everyone else was most likely at the secondary staging up the road. Driving out towards the FTR I saw the ops chief in his truck. Rolling down my window I waved to get his attention.

"Looks like your boys had some fun today," he said over the roar of the wind as he stared out the front windshield. His tired expression didn't seem to match the statement about our ignition crew. "Let's regroup at the equipment staging," he said, referring to a place down the road where the dozers had cleared an area for the heavy equipment to park.

There we reunited with the four firefighters on the ignition crew and listened to their excitement about what had happened. I was pumped to hear that they got that experience but was relieved they were safe.

As afternoon wore on, temperatures decreased, the wind calmed, and the fire began to relax its pace. But the damage was done. After we were released to go back to camp, we watched a group of dozers crawl into the forest towards the orange glow of the fire which was highlighted by the setting sun. Several firefighters from other crews followed the dozers into the forest and were engulfed out of view.

A firefighter walked up to us with shovel in hand. "Which way is the fire?" he asked.

I pointed towards the red haze above the treeline. He nodded his head, turned on his headlamp, and headed in that direction.

"Man, that's cool," said Jordan.

"That will be us soon buddy," I replied.

In less than two hours, the fire had doubled in size. Fortunately for everyone, the wind event occurred at the end of the day and the cold spring air was enough to help stop the fire just one kilometre from Nordegg.

By morning, the fire was contained at 253 hectares. This time we were lucky. If that wind event happened at peak burn and not at the end of the day it would have been a different outcome. To be at the mercy of your environment is a humbling experience. An experience that we would learn from to battle the trials to come.

3

DITCH BEERS

"Don't be silly. There're no sasquatches in Alberta. They live on the West Coast."
– Sam "Dad" Hetherington

We finished our fourth day on the line and since the blow-up of the blaze two days prior, the fire had remained calm, and the cool spring weather helped keep the Fish Lake Fire from growing. Temperatures remained in the single digits and with help from a sprinkle of rain, the fire was easy to hold. Even though the perimeter was contained, the work of digging up and finding the hidden heat remained constant and for the rookies that had never worked in the forest before, fatigue was catching up to them. I added to their workload by pushing them as hard as they physically could because of the pressure I felt from wanting to succeed and to build a good reputation for our crew. With the spotlight on the new Unit program our effectiveness was being judged and I knew how important our first impression needed to be.

We left the fire and began our 30-minute drive back to base. As Alpha truck came to a stop at the junction of the Forestry Trunk Road and the main highway leading back to camp, Dad, who was driving, scanned his mirrors making sure the other three Chevy 3500s were still in sight. The 16-hour days were starting to wear down some of the rookies who had yet to get their bush legs under them. Dad was no different and was grateful that this day was almost over.

Dad quickly appreciated being in Alpha truck which always led our convoy of four vehicles. I rode shotgun in Alpha and with me being the conduit who relayed instructions, travelling in the lead was the obvious choice. Dad enjoyed hearing first-hand the phone conversations I had with other resources and my thoughts about our upcoming objectives.

Another perk of being in Alpha was we never had to eat the dust kicked up by the other trucks on gravel roads. At times, such dust became so thick it would hinder the following drivers' views to only a metre, making it more dangerous for inexperienced drivers. In fact, driving on unmaintained roads involved a steep learning curve for most people who came from big cities.

Delta truck, driven by Andrew Fulthorpe, had snuck into second from third since leaving the Fish Lake Fire. Echo truck, driven by Gary Thirnbeck, an Edmonton-born beast with unnatural strength, took its usual spot at the rear of the convoy, herding the other trucks like a sheepdog to keep on pace with Alpha.

While en route to camp, a non-familiar voice cut in over our truck radio, announcing that: "All crews are to report to the warehouse for a meeting when they get back to camp."

"What do you think the meeting's about?" Dad asked me.

"Probably someone was caught drinking in camp and is getting kicked off the fire. Happens more than you think," I said.

"Uh, we are not allowed to drink after hours?" he asked, the blood draining from his face as he ran his right hand through his curly blond hair.

"Common sense. We are staying at a government camp. Why?" I asked moving my gaze from the road to stare at him, waiting for an answer I did not want to hear.

He could feel my icy stare and after a few awkward moments he knew he had to tell me the truth. Throughout my wildfire career, I figured out that if I deadpan stare at someone after I ask them a question or after they ask something I thought was simple, they would eventually respond appropriately.

"Um, yeah, I went for a walk last night and found a six-pack of beer in the ditch," he confessed. "I brought them back to camp." I continued to stare at him in silence. "I also shared them with Matt and Couse. I didn't know it wasn't allowed."

I held in my laughter and looked at his leader James in the back seat who was smiling ear to ear. "So, let me get this straight," I said, then paused to take in a calming breath. "You went for a walk, found beer in the ditch, thought it was your lucky day, and then drank the ditch beer?"

After hearing Dad's somewhat lame confession, I knew the upcoming meeting could not be about my crew. If it was, it would – or at least it should – have been brought up with me in private before it was addressed to the whole camp. If for some small chance it was about Dad, I was confident I would be able to play to the rookie's ignorance as an excuse. That was something Dad didn't know.

"Don't worry Dad. I'll make sure you get paid for the time you were on the fire," I said unemotionally while moving my stare away from him and back to the road.

"I can't believe you drank ditch beers," James chirped from the back seat. Dad drove the rest of the way back to camp in silence, continuing to nervously run his hand through his wavy blond hair.

We parked our trucks in the usual spot by the bottom of the hill where our tents were set up. It was commonplace for firefighters to stay in tents for weeks at a time while fighting a campaign fire and this was no different. It was springtime in Northern Canada and the nighttime lows consistently dropped below freezing. The cold temperatures did not extinguish the fires, though. Instead, they lulled the fire to sleep until the warmth of the sun would wake the crimson beast to begin its daily search for fuel to consume.

During the previous fire season, Farley and I had spent over 50 nights in our tents and had come prepared to deal with the cold nights. Down-filled sleeping bags rated for -10°C, toques, and wool socks helped make the cool nights bearable. Most of the crew, especially the firefighters who lived in the south, only brought thin sleeping bags like the ones you find in your local Canadian Tire discount bin. The frosty temperatures keeping them awake at night only added to their fatigue. Any mention of being cold fell on my deaf ears since they had all been told to bring warm clothes and sleeping bags during their rookie training.

It wasn't just the rookies that were cold at night. Warhammer, who had forgot his personal bag in the laundry room back in Manning, had been using his backup Nomex uniform as a towel. He then used the same wet uniform as a blanket in the frigid temperatures. Yet he never complained. Unlike the rookies, Warhammer held a special kind of inner strength that wouldn't let something as trivial as the cold wear him down.

Dad didn't mind the cool nights either as he was one of the rookies with enough forethought to bring appropriate camping gear. He was tired, certainly, but happy to be part of this new team of valuable firefighters. His happiness, though, was but an afterthought to the meeting that we were now summoned to.

We got out of our trucks and mixed in with the rest of the firefighters that had gathered around the Incident Commander who stood in front of the warehouse. We were easy to distinguish from the others with each of us wearing our two-piece uniforms of dark-green pants and yellow shirts. Most of the other firefighting crews wore yellow coveralls.

That was the first year that firefighters who didn't wear a white hard hat and weren't full-time employees, were issued two-piece uniforms. Before this, the only option was to wear a one-piece yellow coverall that made us look like convicts working off community service duty. Most of the experienced firefighters, me especially, loved the new uniforms. In the past while working in other provinces I had to deal with being treated like a contractor simply because my government was behind the curve on providing us professional-looking uniforms. Because of this I banned anyone on my crew from wearing coveralls, especially when we were in the public eye. Wanting to look the part I made sure my crew adopted the new uniforms that still seemed foreign to the other crews.

"Hello everyone," shouted the IC, causing the dozen or so conversations between the firefighters to stop. "First of all, I would like to thank all of you for your hard work. We have stopped the fire and saved the town of Nordegg. You should all be very proud, and this is a story you will be able to tell your kids one day." The crowd clapped and cheered. "As a reward," he continued, "you can all start at 07:00 tomorrow."

Several more cheers sounded from the group with none louder than Adam. I almost laughed at how excited some people were to hear the news. Most of the leaders failed to see how an extra hour in the morning could make any difference to our day but that extra hour of sleep was still appreciated. We were proud of the work we had accomplished and were grateful for any reward.

"The reason I gathered you all here is because it's our warehouse leader's birthday and we got cake for everyone," said the IC.

Dad, hearing he was in the clear, let out a long breath like he had been holding it in for the last 20 minutes. Looking behind him, he saw James and me smiling and shaking our heads. James mouthed the words "ditch beers" to him before Dad turned back, embarrassed that the two of us had had a joke at his expense. The rest of the crew congregated around the cake to celebrate the end of the day before returning tired and sore to their tents. (By the way Dad, I know you didn't find those beers in the ditch.)

From that day forward the fire did not grow and the job became a mop-up show, the search and destroy of the remaining heat on the fire. Three days later, the four remaining rookies – Nate Thompson, Courtney Stevenson, Colin Mckinnon, and Jesse Hoevenaars – arrived at the fire after finishing the last rookie training camp of the season. They brought our gear trailer hitched to the back of the truck as well as Warhammer's personal bag that contained a change of clothes, a towel, pillow, and his sleeping bag. Those four rookies missed the flames but had like the rest of us, learned several important extinguishment strategies such as helicopter operations, hose-lay systems, chainsaw bucking and tree falling, and how to utilize the heavy equipment such as dozers, excavators, skidders, and nodwells.

During the mop-up stage, helicopter bucketing operations became a favourite task for my crew. Up to 12 helicopters at a time were used for bucket support, medivac, mapping in the morning, and reconnaissance. This included five Bell 212s, two 214s, one 205, and four A-Stars. This provided the rookies with experience calling in bucket drops which would require them to radio in a helicopter with a full bucket of water to drop on a hot spot that they had dug up and flagged.

On our last day on the line, I was able to sweet-talk the air operations supervisor into giving everyone on my crew a flight over the fire. For most of the rookies, it was their first time in a helicopter and an unforgettable reward for their hard work.

We spent a total of 162 hours on the Fish Lake Fire over a 12-day period. We were then released and made the eight-hour drive back north to our home base in Manning. We saw the fire from its final blow-up stage to almost complete extinguishment. And yeah, I was happy with the efforts we put forward on our first fire as a crew.

We were tested and passed our first major incident. Valuable lessons were learned, and most importantly, no one was hurt. I could see our potential and knew in my heart what we could become with the proper training and leadership.

We came to the fire as strangers and returned home as a team. The Peace River Unit Crew was here.

4

THE BOMBER BOYS OF MANNING

"Hit the highway and head north."
– Angry farming dad

Home was the Manning Air Tanker Base (MATB) in Northern Alberta. The base sits in the middle of the small farming community of Manning, about 100 kilometres north of the town of Peace River. The 700 people that live in and around the town year-round consist mainly of farmers that make their living growing crops of oats, hay, and canola. There is no compelling reason for anyone to go out of their way to visit this sleepy town named after the former Alberta Premier Ernest Manning, except to stop for gas along the main highway to the Northwest Territories.

For the previous eight summers I called this town my home, and after every winter I looked forward to coming back to my hidden paradise. Each spring the MATB would open, and the firefighters and support staff would slowly filter in. By summer, more than 60 people would be there including cooks and cleaning staff.

The Notikewin River, a tributary of the Peace River, flows through town south of the base. Here firefighters target large pike and walleye in their downtimes. Miles of farmland surrounds the base giving the air-tankers and helicopters plenty of space to land on either the large tarmac or on one of the three helipads.

The MATB was made up of several buildings: living quarters for the firefighters plus several ATCO trailers for the overflow of new firefighters, a rec building with a pool table and a big screen television, a kitchen, office, gym, and a larger building where the air-tanker staff would work.

All the buildings had the same look – white aluminum siding covered by pine-green aluminum roofs. The base had been there for decades and had housed hundreds of firefighters throughout its lifespan. Now with the influx of 20 more people, the district had replaced some of the older buildings including the kitchen and sleeper trailers.

The support staff that looked after the firefighters represented the lowest-bidding contractors. Before this season, Walt and Debbie, an older couple, ran the base. Walt was a great camp boss and Debbie was the head cook – and she loved making homemade meals for the crew. The two had been there since before I was a rookie and looked after the firefighters as if we were their own children. Walt was diagnosed with cancer, and they retired to their farm in Saskatchewan. I missed them both.

The rules for living on the base seemed strict on paper, but with the Peace River Office being over an hour's drive to the south, it was hard for anyone to enforce the rules. It was up to the senior firefighters to keep the crews in line and to make sure they did not get up to anything that would reflect poorly on the Alberta

Government. It was supposed to be a "dry camp," meaning no alcohol, but when you get 50 or so young men and women in the middle of nowhere, away from friends and families, *that* is a hard rule to enforce. If during working hours the rules were followed and no bad news got back to the Office, the crews for the most part were trusted. And everyone knew at least a few cautionary tales about past firefighters who had been a little too cavalier with the rules and were not rehired back – these stories kept most firefighters pretty much in check. For sure, seeing past firefighters and friends that could no longer work beside us kept senior firefighters looking out for everyone else.

Being located so far north, the summer days were long making for some exhausting workdays. During the peak of summer, there were 19 hours of daylight, leaving only five hours of twilight in the absence of direct sun. With the flat landscape stretching farther than one could see, sunrises and sunsets would fill the sky with bright hues of red, pink, orange, and purple. At times, several storms at once could be seen in their entirety making Manning seem as if it was perched on top of the world. During the cool nights in early spring and late autumn, the Northern Lights would sweep from one end of the horizon to the other, dancing across the night sky in shades of emerald greens, topaz blues, and ruby reds.

Manning also happens to be in the middle of one of the main bird migratory routes in North America, with hundreds of species filling the local airspace. Birding would become a favourite pastime for me and others on the crew.

Living on base certainly had its benefits. I got to bond with the crew and had free room and board. However, to keep my sanity I had to get off base after work hours to separate myself from the job. After the workday was done, I would patrol the back roads either in "Baby Blue," my small Chevy S-10 truck, or in Jordan's green Subaru Outback. Here I'd seemingly drop into the live version of a

Discovery Channel show featuring northern harriers soaring over crops looking for voles, or waterfowl making use of the dugouts on their way to their northern breeding grounds.

Firefighters who arrived every year were called "Bomber Boys" by the locals. Only about 10 percent of the firefighters were female. And with so many single men on base many would go into town in search of some "lonely farmer's daughter." That would usually start a feud between the local farming men and the Bomber Boys of summer.

The MATB would be where we would sleep, eat, exercise, work, and play, until each crew would disband at the end of the fire season. That's when the base was shut down, usually in the second week of September.

My girlfriend Cavelle, who I had met in college, wanted to be close to me for the summer, so she applied for jobs in the district. She got accepted as a Lookout at the Notikewin Lookout tower.

The Peace River District had 14 lookout towers that were each occupied by someone that, depending on the likelihood of fire, would sit in a six-square-metre octagon copula perched on steel trusses 25 to 30 metres in the air. Throughout my time in the Peace, I made personal connections with most of the tower people that would call the remote boreal wilderness their home. For four to six months at a time, these people would stay at their towers and vigilantly look for fires. Half the lookouts in the district were only accessible by helicopter and sometimes my crew would represent the only human contact they would have for weeks at a time.

It is a beautiful and romantic notion to think about living in the Canadian solitude but as I discovered, isolation can cause unique changes to the mind. Or maybe people that like to live alone already possess these "changes," they had to be different in the first place to want to have a life so far removed from society. Whatever the case, technology would slowly replace the need for these lookouts and over time the people that staffed the lookout towers of Northern Alberta would become a part of the past.

I didn't know that Cavelle had applied for work up north and was pleasantly surprised to hear the news. The only problem was that the Notikewin Lookout was one of the fly-in locations. For fuck sakes. My first attempt to see her failed because I was turned around from a washed-out road and angry bears. Not wanting to be eaten by large carnivores or drowned in the river crossings I came up with a simpler plan. With help from a local in town, I was introduced to the oil pilots that took daily flights to check on the well sites.[4] By trading premium bottles of whiskey, I was able to hitch rides to and from my beautiful girlfriend's tower.

It was an enjoyable way to spend my days off. Being alone in the forest with Cavelle and her 80-pound dog Diesel was relaxing and gave me a break from the Unit life. During the days we would sit in the lookout 25 metres above the ground looking for fires, sharing an experience that few couples had. At night we would cuddle around a campfire and tell stories about our lives. I was completely in love with her.

But once back at base my focus shifted solely to the Unit. I was pleased with the effort we had put forward on the Fish Lake Fire and was excited for our potential. Even though the fire was stopped and the town of Nordegg was saved, I knew how much better the crew could be. We were put at the heel of the fire, and I needed a crew that would be sent to the frontlines. The learning curve was steep. It was the first time most of my crew had fought a wildfire. The sub-leaders, all former Helitack firefighters, had not been exposed to many large fires and their new roles as leaders

---

4 Well site – A cleared area of land usually 100m^2 that had been drilled for oil or gas.

were just as important to figure out as was fighting this new heavyweight opponent.

The only way to get better until the next fire was to train and prepare for the next deployment. I did not just have the task of training the rookies on how to become proper firefighters; I also had to train my sub-leaders on how to lead their crews. This was new to me; I had never even worked on a Unit crew never mind leading one. With my experience, I was confident that I was the best person for the job, but I knew there was going to be a lot of trial and error over the foreseeable future until a foundation and routine could be established. Seeing that I wasn't going to get any help from my former Helitack buddies, it looked like I was on my own.

Before the rookies came to Peace River, they completed a two-week wildland firefighter training course at the Hinton Training Centre. They acquired the basics but the majority of what they needed to learn would come through on-the-job training. We would spend our time learning the ins and outs of the equipment such as chainsaws, pumps, and hand tools. An hour was set aside every day for fitness training which became a favourite time of the day for DK. Like Jordan, DK also studied physical education and excelled at making activities to challenge the crew physically.

Occasionally we would go into town and use the local pool for fitness training. There we learned that all four men on Echo crew could not swim. From there on, they would be the terrestrial crew and kept away from any deep water.

When not training, we would be sent to do project work like cleaning up the areas around the lookout towers, setting up electric fences to keep the bears out of camps and towers, and general maintenance work.

The amount of paperwork that is needed while working for the government can be staggering and it is something that is not taught to the leaders during their 10-day leader course in Hinton. I would also have to spend time with each of the five sub-leaders and train them on the daily paperwork such as timesheets, training logs, and crew appraisals.

The beginning of the summer remained cool and wet throughout Alberta. Daily weather can be somewhat predictable but long-term trends associated with weather phenomena such as El Nino and El Nina are fickle and cannot be accurately predicted. All said, the extended periods of hot and dry weather that was promised by the weather forecasters never came.

As much we wanted to get back to the line and fight another large fire, the opportunity relied on the hot, dry weather that never lasted long. We did fight three small lightning-strike wildfires close to base, but the only real challenge in those fires was teaching the recruits how to traverse through muskeg and deal with mosquitos. I and some of the leaders went to a small powerline fire that took a couple hours of our time but besides that, the early part of summer was uneventful.

Because the fire season was so incredibly wet and cool and the firefighters across the province were not needed for emergency services, for the first time in several years the Alberta Wildfire Olympics were held in Hinton. Crews would compete in 10 events such as fire starting, fitness challenges, and chainsaw skills. Farley, Soso, Dad, and DK strongly represented the Peace River Unit over three days of competition where they placed 8th out of 18 teams.

It would not be until the end of June when we were needed again to help the province of Alberta. This time the rain came with vengeance creating a different type of natural disaster. Instead of pulling water towards our objectives, we would be pushing it away.

5

REDWOOD MEADOWS

"OooOOooo."
– Albert Cooper, Peace River Unit
crew coordinator 2015 - 2018

On the morning of June 21, we reported to the Calgary Wildfire headquarters for our upcoming assignment. By the time I got out of the truck and ran to the front door of the old building, I was soaked from the torrential downpour.

"Holy shit," I said to myself as I wiped the rain from my face. My five sub-leaders did the same while shaking the water from their bodies. The six of us along with the rest of the Unit looked mismatched in our attempts to stay dry and warm. Grey wool Stanfield's that kept us surprisingly warm even when wet became a favourite for the crew. Over the warm wool were rain jackets of various colours that mixed over our green Nomex pants.

"Oh man. My socks are soaked," DK said to James as we went for our district briefing. Being the two youngest leaders on the Unit, the two had developed a strong bond since the start of the season.

The rain had come down in record numbers over the previous 48 hours. Over 220 millimetres had fallen in the Canmore area and over 335 millimetres in High River. And there was no break in sight. The high levels of snowfall that melted in the spring meant the ground was already saturated so this force of precipitation plus the state of groundwater made for a deadly mix. No doubt we would be needed for flood relief.

We knew this before the briefing. We had arrived in Calgary the previous night and the destruction from the overflowing Bow River that ran through the city was heartbreaking. I had previously lived in Calgary and knew how devastating the scene was.

No one on our crew had ever fought a flood before. For that matter, it was doubtful that anyone in Alberta Wildfire had, so we didn't really have any clue as to what was going to be expected of us. Since the fire in Nordegg over a month earlier, the fire season had been slow and the chance to do some meaningful work was exciting.

We were instructed to go to the community of Redwood Meadows and meet up with the local fire department to do whatever they needed.

I knew that fighting a flood was rare. So I told the Unit that this would most likely be a once in a lifetime opportunity and not to take this time in their lives for granted. We were hired on as firefighters, but now we had a chance to help others without extinguishing flames. Instead of bringing water to danger we would be forcing it away. As Jordan would later describe it in an interview, firefighters are professionals in hydrology, the movement of water in relation to land. Transporting water from one area to another was our speciality and this unique skill was about to come in handy.

The 20 of us piled into our five trucks and followed a forest ranger that worked in Calgary. We headed southwest towards our next objective.

The wiper blades set to their fastest cleared the windshield just enough for Dad to see where we were going. He leaned into the steering wheel hoping that extra couple inches would help increase his visibility.

Almost every side road that lined the main highway was washed out or completely covered in water. The bridge that spanned the pale blue stream just south of Bragg Creek was now a raging torrent turned brown with sand and silt runoff. Large trees and debris mixed in with the turbulent rapids flowing downstream to some unknown destination.

Driving for 45 minutes through the pouring rain we finally arrived at a Redwood Meadows fire hall. During the initial stages of the storm, the town had lost power and the residents had been asked to leave for their own safety. The Elbow River that flowed through town raged against the riverbank. It was on the verge of spilling onto the streets and washing out the town of over 1,000 people.

An orange excavator sat abandoned on a high point by the Elbow River. It had earlier been used in building up a barrier at a bend in the river made of large black rocks piled 20 metres in length and 3 metres high. Large cement blocks were lined in front of the rocks to makeshift the rock wall reinforcement. The river rushed less than a foot from the wall foundation. With the ferocity of the raging waters, the wall did not look like it would be enough to hold if the Elbow were to test its strength.

There wasn't a clear objective when we arrived, so we waited in the fire hall while the ranger sought work for us. While he was seeking an objective, a local air attack officer (AAO) – the person who sits in the front passenger seat (not the pilot) of the Bird Dogs that lead the air tankers on fire runs – asked if he could get some help. He requested a sub-crew and a pump to help de-flood a nearby basement. I took Alpha crew and left the fire hall. Once again, I got soaked having to run through the thick sheets of precipitation to get to our truck.

We followed this older man several minutes through the flooded community and stopped at the side of a cul-de-sac. Brown water flowed over the road and funnelled into small whirlpools over the drains as the water tried in vain to empty from the streets.

An RCMP car was parked on the other side of the street from us. Since the evacuation order had been put in place, the police went door to door checking for anyone that had not yet got the message. I gave a slight head nod to the cop who responded with the same. He continued to the next house paying us scant attention. Except for him, the streets were empty of people and vehicles.

The Alberta Wildfire workers had been given permission to stay because they were an emergency service, and this local AAO seemed to share the same autonomy. Upon arrival at the flooded house, James, Dad, Courtney, and I found that it was in fact the home of the AAO.

The middle-aged man with grey hair and diamond-studded earrings introduced us to his wife who had been salvaging what she could from their basement. She was happy to see that help had come and more cheerful than I expected for someone that was knee-deep in a natural disaster.

"Thank you so much for coming over. Don't worry about your boots, honey," she said. The carpets in the main doorway were soaked and muddy and would have to be replaced.

Walking down the wooden stairs to the basement, I was surprised how high the water was. The brown water was over a metre high and must have been seeping into the house from the ground. The water outside remained on the roads and had yet to rise above the sidewalks.

Most of the basement had been cleared out except for the larger appliances like the washing machine and dryer. Loose articles of clothing and random items floated in the dark water that covered the cemented basement floor.

We had brought our Floto-pump knowing it would be more suited for this task than the other two options of the Mark-3 and the mini-Mark. This was because the Floto did not need a suction hose or a foot valve to get its prime. The water intake was built into the bottom of the pump pulling the water in with centrifugal force, allowing it to suck up water in shallower depths compared to other pumps that needed a foot valve to be fully submerged. It also had a self-contained gas tank so there was no need to find a dry, safe place to put a jerry can.

James opened the basement window and fed a length of 1 ½" hose out the window to Dad who then ran the other end to the street. James connected the hose to the pump and after a couple of pulls from the rip cord and a tweak of the throttle, the Floto roared to life. Black smoke billowed from the exhaust and the revving from the engine was amplified in the closed structure. Pulling down our green peltors from our helmets to muffle the noise, we watched as the pump began to push water through the hose and out onto the flooded street.

It wasn't long before the water level began to drop; however, with only a small window for ventilation, the excess exhaust began to filter into the upper levels of the house. Within 20 minutes, the dark water that was thigh high was now under our knees and the light-coloured concrete floor was visible. Even with the doors and windows wide open, the smell from the exhaust was prevalent and becoming more of a problem than the flooding in the basement.

This didn't seem to bother the couple that were fixated on removing the water that had invaded their home. I knew that whatever damage the basement had sustained was not going to get better. So I talked the AAO into focusing his attention to the fact that the exhaust was going to ruin other parts of the house.

After a brief conversation with his wife, he agreed to shut down the pump and let the flooding in the basement do what it was going to do. Without an electric pump and a generator to run it because the town power was out, the water would have to stay until it drained on its own.

The destructive smoke from the pump exhaust was a problem – and a valuable learning experience for the work ahead.

We packed our gear, said goodbye and good luck to the older forestry couple and returned to the fire hall. While we were attempting to pump out that house, the rest of the Unit was assigned to help reinforce the rock barrier along the river with sandbags. Rain gear did little to protect them from the heavy rain still pelting the region. Without complaint they eagerly stacked sandbag after sandbag along the shore of the rising Elbow River. With Stanfield's heavy from rain and leather boots filled with water, they happily worked until dark.

By morning, the rain that had been coming down for 70 hours stopped and the sun was free from the clouds that had kept it hidden. Water levels subsided, and the strength of the rock wall was not tested. But the effect of the flood was widespread. Thirty-two states of local emergency were declared and over 100,000 people were displaced from their homes throughout Southern Alberta.

But our experience with flood waters was just beginning. Our next assignment: The community of Black Diamond, a town that had felt the full force of the flood and was left overwhelmed in its aftermath.

6

THE YELLOW SHIRT MARINES

"It was phenomenal. They worked hard, nose down, flat up, and got it done. They were amazing."
– High River Fire Chief Campbell

During our drive from Manning to Calgary, a wildfire ranger from the town of Black Diamond was at his home getting ready to leave for work. Black Diamond was one of the smaller communities south of Calgary and 2,600 people called this flood plain their home.

From his living room window, the ranger could see that the Sheep River had rose to an unnatural level and had turned brown from the fast-flowing water mixing with dirt. For the last two days, it had been raining harder than he could remember and like all wildland firefighters, the deluge signalled to him that the fire season, at least in the Calgary District, would be on hold for the time being. June monsoon was the term for the early summer rains that were common in Alberta, but there was nothing common about the amount of precipitation that was now falling.

The ranger ran from the front door to his truck and was soaked in that brief time with heavy rain. He left his driveway and swerved around large pools of water that accumulated on the road. Turning from his street the road ahead disappeared under dark rushing water.

Water almost a metre high was rushing over the road, pushing into the gas station and strip mall. The Sheep River that flowed through the north of town had risen over its banks and flooded the northwest section of the town. He now knew he had more important tasks than to show up to work that morning. He phoned his duty officer and got permission to do what he could for his community.

Instead of driving to his office in Calgary, he veered from his Monday to Friday route and went to the town hall to register with the local volunteer committee. There he informed the staff of his role with Alberta Wildfire and the services they could offer, including firefighters and equipment. Upon hearing this, the town's fire chief accepted the help without hesitation. It was an opportunity to get emergency responders who were trained in dealing with natural disasters and there was no logical reason why such help should be turned away.

The ranger's request to Alberta Wildfire was granted, and the allocation of resources was given to help. Eleven firefighters would show up the next day with my Unit arriving two days later to begin our new assignment: tackling the Black Diamond flood. There in his hometown, the ranger would become the official liaison between Alberta Wildfire and the Black Diamond volunteer committee.

When we arrived in Black Diamond on the morning of June 22, the muddy waters that had been flowing several feet over the road and into the community had receded back into the Sheep River. Debris that washed from upstream and onto the streets were scattered everywhere. Chunks of wood, natural and manufactured materials that ranged from small sticks to large pieces of fencing lay in the most curious resting places. All this debris as well as any surface that was once submerged in water was covered in light brown sediment that had saturated into the flood waters.

As the 20 of us exited our trucks at the town fire hall, the thick smell of mud filled our nostrils. While standing on the sidewalk we surveyed the area around us and wondered what our objective was going to be. Seeing earlier flood relief on the TV news, the only work I could imagine us doing was sandbagging against the high-water mark or helping with evacuations. Since the Sheep River was now back to a reasonable level and the flood waters drained from the streets, those two tasks could not be what we were here for.

We were met by the ranger and the volunteer fire department chief on the sidewalk in front of the main entrance to the hall.

The two middle-aged men explained that there were over 70 homes that had been affected by the flooding. Many had their basements filled with water close to a metre high for more than 24 hours. Anything that was not waterproof was destroyed. Not just from the water but from the silt and sewage that accompanied it. Major appliances, furniture, clothes, children's toys, drywall, carpet, priceless photo albums, the list went on.

While the two continued to describe the situation, a four-person Helitack crew and a seven-person Rappel crew joined our group to be briefed on the day's assignment. The Rappel crew was led by Rob Munday, a tall slender Australian firefighter that I had worked with Down Under during the Black Saturday Bush Fires that raged there in February 2009. Outside of work, Munday and I most likely wouldn't be friends, but the bond that is created from this job transcends the norm and friendships can be made between the most opposite of personalities.

Those 11 other firefighters arrived at Black Diamond the previous day and had been pumping out basements in the southwest of town where the water levels still hadn't dropped back to normal. During the briefing, they were instructed to continue work on the houses they had started to un-flood.

We would be assigned to the north end of town. "There's a lot of people that need your help. Focus on the houses then work on the streets," said the ranger. "Garbage trucks are being brought in by volunteer organizations and will be coming in to take away whatever you put on the streets."

"No worries. We'll do our best," I replied.

After the briefing we drove three blocks to the middle of where the flooding had occurred and calculated the damage. Debris of all kinds were strewn around us. A handful of residents roamed their front yards while others sat on their decks with blank expressions.

"So, what's the plan here?" Jordan asked as the sub-leaders and I gathered for a meeting.

"Let's just start by going door to door and seeing who needs help," I said. "Remember that these people will be emotional, and it's our job to help them however we can. It doesn't matter what they ask because it will be important to them."

"Start small, get big," Jordan said, referring to the motto we had acquired after the Fish Lake Fire.

From there I divided up the streets into three working areas with one or two crews assigned to each. Dad's extended family lived in one of the houses that was in need, so I gave him the freedom to go help on that property.

Most residents had not returned from wherever they had sought refuge. Most of the ones that had returned seemed confused at what they should be doing. What happened to them and how they were dealing with it was unique to each of them. In my experience, when people are personally affected by a natural disaster their emotions directly coincide with the classic five stages of grief: denial, anger, bargaining, depression, and finally, acceptance. The loss here to residents in Black Diamond was out of anyone's control and it reminded me how much this situation was like a wildfire: when disruption to residents' lives from a natural disaster is new – the grief stage of acceptance is so far from their grasps as to be completely off the radar.

Scanning down the dirty, silt covered street, I saw a blonde, middle-aged woman sitting on the steps in front of her door. Her hand shook as she drew in a long breath of smoke from her cigarette and washed it down with what I could only reason was a stiff drink.

Walking passed two empty houses I crossed her mud-covered lawn and sat beside her. "Hello. I'm Harold from Alberta Wildfire," I said. "How you holding up?"

The lady took another swig from her drink then put the glass down on the cement steps. "I don't know what to do. Everything in my basement is ruined," she said. Her eyes welled up and a tear from each eye rolled down her puffy cheeks.

"I know it is hard to imagine right now but this will get better. I have a crew here and we will help however we can. Show me what's damaged and we can take it away for you."

She wiped the tears from her cheeks, put out her cigarette and threw the butt in the pile that had acuminated beside her steps. "Thank you so much but I'm not sure what you can do," she said.

"Let's go take a look and see," I replied.

I followed her into her home and down the wooden stairs that led to the basement. Over 10 centimetres of water remained on the surface of the concrete floor. The basement had cement walls with unfinished two-by-four frames intended to hold electrical wire and insulation. A dark line on the walls over a metre from the cement floor showed how high the water had been. Discarded clothes and personal items were spread all over. A soiled couch and matching chair as well as washing and drying machines were among the larger items in the basement. She told me how she was going through a divorce and recently moved, adding to her stress, but because of the move she had yet to acquire a vast number of personal items. Her family and friends were unable to help, and she was alone.

"No worries. We'll have all of this taken care of in no time," I said. Over the radio, I called Echo crew and within minutes anything she wanted taken away was being carried up the stairs and thrown into a pile on her driveway. Gary and the Warhammer's unnatural strength became an asset when moving the heavy appliances. Farley and Gadget[5] didn't let their smaller frames slow them down and kept pace with the two heavyweights on their crew. In under an hour, the five of us had completely gutted her basement and piled the contents on the corner of the driveway. When garbage trucks arrived, we moved the pile of trash onto the truck beds to be brought to a nearby dump.

---

5 On the first day of the Fish Lake Fire, Ian Megega showed up with a backpack full of rope and gear that he thought would be useful on the line. He was henceforth known as Gadget.

Seeing her basement emptied and driveway cleared, she threw her arms around me and cried on my shoulder. "Thank you so much," she said between sobs. After the long hug, I wished her luck and went to the next house to see what we could do.

No matter what we did, it was impossible to stay clean. Just walking around, never mind hauling and moving silt and mud-covered items, our uniforms quickly became covered in a layer of brownish grey sludge. The dirt stood out on our dark-green pants and bright yellow Nomex shirts. Most of the crew, however, had already become used to being covered in dirt and ash so this flood sludge did little to affect their work performance.

The only interruption to the steady flow of work was when you heard fabric tear followed by a loud cuss. The stitching in the pants was weak often causing the crotch or ass to blow apart. During the Fish Lake Fire, most firefighters had ripped their pants while working the line. After the fire, we stitched them ourselves since getting replacements from the warehouse was not worth the bitching we'd get by the warehouse staff. The office staff couldn't understand why all the firefighters were blowing out their pants since they wore the same ones as us. It took a couple of years for the office staff to understand that the firefighters in the field do hard physical work, so reinforced stitching eventually became mandatory on pant orders. But until that epiphany, pants with the crotch or ass blown out were a common sight among crews that didn't have the necessary sewing skills.

The benefit of this, though, was the breeze that would enter our pants and cool the intense chaffing that the Nomex caused on our inner thighs and butt checks. Still to this day, hair doesn't grow in those areas from the years of chaffing. Pro tip: Gold Bond medicated powder is essential for the first few weeks on the line.

The neighbourhood we were working in held an older population and they were incredibly grateful for the help. Seeing the impact we were making only fuelled our efforts. We went from house to house helping whoever needed it until the end of the workday when we were told to stop. Progress went quickly and gained momentum as we figured out how to help. During the flood we had to get into deep squats and unnatural positions to lift heavy items. And sure enough, by the end of the first day over half the crew had once again blown open their pants in the ass and crotch.

When we arrived the following day, the street that we had been working on was alive with people. Either the residents had decided to come home on their own accord to check on their houses or word spread that help was available. Having the homeowners there allowed us to go back to the houses that we had missed and continue the process of cleaning out any damaged property. Not just that, we also shovelled any dirt or mud that had accumulated on the roads and sidewalks.

Dozens of truckloads of garbage continuously filtered out of Black Diamond to the closest landfill. We worked at such a pace that the trucks could not keep up with the piles accumulating on the driveways and streets.

I began to see a different side of my crew than the one I witnessed on the Fish Lake Fire. Unlike the town of Nordegg, we could interact with the people we were helping here, and this pushed the crew to go beyond their limits for those in need. I was impressed that no one complained about having to lift heavy objects from basements all day. Or the smell. The muddy odour that filled the air was becoming stale with hints of sewage mixed in. We were able to don much needed dust masks and rubber gloves that had been donated to the town hall. When working with any garbage that seemed to be moulding or contaminated with sewage, the extra protection was appreciated.

Each person on the crew had their own strength when dealing with the public. Jordan was personable and a major help talking to the homeowners and seamlessly finding work for the other sub-crews. The Fulthorpe brothers were beasts when it came to utilizing their physical strength. Colin McKinnon's 6'10" avatar was a blessing for moving objects – no wonder he was jokingly called the "Forklift."

For the second straight day we crushed the streets of Black Diamond. Each crew moved from house to house gutting the basement from any evidence that there was a flood. DK and his Charlie crew were unfortunate to find a hoarder and for a second straight day were contained there, bringing out the thousands of items in that wrecked basement. The old lady who lived there and her reluctance to throw anything out consumed much of their time

After each house was cleaned out, the homeowners would show their appreciation in a variety of ways. Some with food and drink. Beers and other alcoholic beverages were the favourite gift but each offer was respectively declined. The best show of appreciation was the handshakes, the hugs, and the happy tears. For experienced firefighters, working with the public and having them directly show their gratitude was something foreign because they spent so much of their time in the forest. It was heartwarming to have people showing thanks in a usually thankless job.

By the third day of cleanup, the local paper headline read "The Marines of Black Diamond," and shared stories about the firefighters working throughout town. Our feverous work ethic with little to no breaks gave us that well-deserved nickname. By the end of the day, all 70 houses and the surrounding streets had been cleaned out.

As we drove out of town for the final time, the residents lined the streets and cheered and clapped in thanks. Adam leaned out the truck window and held up a Canadian flag while he cheered along with the town.

Our exposure to the flooding had shown us how damaging this type of natural disaster can be and once again how much value there is when 20 people work as a team. I was like a proud father to my crew. Seeing the young men and women grow together when it was needed most kept a secret smile behind my usually stoic face. What I had imagined and set out for at the beginning of the season was happening. The Peace River Unit was becoming an unstoppable force. A force that would be needed for the days to come.

7

THE BRIDGE

"It was jammed up there like a bird's nest."
– Todd Lynch, Wildfire Assessor

"Any ideas how we're going to deal with that," I said to my sub-leaders. We stared at our task of unblocking the river crossing into the town of High River which was built in the flood plains south of Calgary. The four-lane bridge that connected the town to the land north was covered in a thick layer of mud. The 200-metre-wide Highwood River below was jammed a third of the way from the north edge to the first pillar that held up the bridge. Logs, wood, and garbage had woven together in the water, collecting more debris as it washed in the jam. Until the river was clear, and the pillar could be inspected, the main road in and out of town would remain closed.

The houses along the riverbank were bereft of their owners. The brown watermark along the houses showed that the river was once above the doorways. Some houses remained intact while others had crumbled into themselves as their foundations were washed away.

"Let's get in there and start pulling it apart," said Jordan.

The muddy river flowed quickly, and loose floating debris added to the log jam with every passing moment. "That water's pretty fast. Not the safest option to be putting people in there," I replied.

"Guess that means Echo crews not helping today," said James. The group chuckled at Warhammer, the Echo sub-leader. Warhammer leaned into the hood of his truck and began rubbing his body on the hood. Looking back at me he smiled with his mouth wide open. It was almost 11:00 so he was probably on his eighth cup of coffee and for the next couple hours he would be unpredictable.

"Dammit Warhammer," I said. He shook his head and stood at attention. The others burst out laughing.

This was our fourth day working in High River and this was the first really challenging objective we were able to tackle. Over the previous three days, we pumped out a soccer field and drove around the northwest section of town identifying any hazards that needed to be cleaned up before the residents could return.

It had been a week and a half since the flood had devastated the town, home to 13,000 people. Half the town was still underwater, and the streets remained empty except for the first responders. Tomorrow would be the first day since the evacuation order that some of the residents were able to come back to town but only if the main road was safe to pass.

How could we dislodge the river jam without going into the water? I did a quick inventory of our gear in my head.

"Let's blast it open with our pumps," I said. "James and Jordo, set up your pumps and work on the jam. Marco and DK, use your pumps to clean the bridge and the streets. Warhammer, take the pike poles to the bridge and see what you can do from above."

"Got it boss," replied Warhammer.

"And Jordo, make sure no one drowns." I said. Jordan was a lifeguard for seven years and his skills helped me feel more at ease around water.

The leaders dispersed and began their assigned tasks. One by one the four pumps roared to life as they sucked water from the river and shot it out through the Hansen nozzles. The smaller debris were easily forced from the jam and the bigger pieces slowly began to float away down river. Two yellow 15-foot pike poles poked at the larger logs from above, loosening their grip on the surrounding flotsam.

On the road above, mud and water flew over the sides of the bridge as water from the hose lines pierced the dried layer of mud, revealing the asphalt hidden underneath.

After three hours, the jam had reduced by half but water from the nozzles stopped moving the thickest of the logs that had woven together from the shore to the pillar. With the smaller debris removed, it revealed that the sandbar was higher than expected and the larger logs rested on the shore. No amount of water being shot at them would move the giant logs.

A local ranger came by to check on our progress. "Looks like it's chainsaw time boys," he said. The ranger strapped on a pair of chaps and along with Marco, the two began cutting the logs from the shoreline. They started with the smaller pieces. Each piece that was cut was thrown into the river to join with the rest of the bird's nest that was slowly unravelling. Gradually they ventured into the water cutting deeper into the jam.

Water roostered into the air behind them as they cut into the submerged wood. As the ranger finished a cut into one of the largest logs a loud snap erupted from under the bridge. The jam suddenly folded into itself loosening the logs that were entangled together. They scrambled to shore as the jam broke apart and floated down river. Cheers erupted over the roar of the pumps from the crew and other responders that had gathered to see the work. The river was clear.

I ran to the top of the bridge and hung out with Farley while watching the rest of the logs peel away from the pillar and float down the brown river.

"Nice work buddy," I said.

"Yeah no doubt. Did you hear about what just happened in the States?" he replied.

"No man. What happened?"

"A Hotshot crew got burned over. Killed all of them but one."

The feeling of accomplishment faded with the news. While we were clearing the bridge in High River on that June 30, 2013, 19 firefighters from the Granite Mountain Hotshots died defending their hometown while on the Yarnell Fire in Arizona. Hearing about the loss sent a wave of emotion through everyone on our Peace River Unit, causing each of us to reflect on our own situation and how you could not possibly put into words how it would feel to lose everyone on your crew except yourself. I could not imagine how it would feel to lose the 19 people that I was responsible for, and it reinforced how much I needed to keep them safe. After the shift, we sent a signed picture of our crew to their base with our deepest condolences to their families.

By dinnertime, we had cleaned the streets and the bridge in our area. Three small dozers arrived. The drivers said they were told to scrape the mud from the streets but couldn't find where they were supposed to go.

"Job's all done here fellas. Thanks anyways," I said. We had completed our task and the town was now ready to let at least some of the residents back in.

We headed back to our tents that were set up in the high school playing field. The high school had been taken over and renamed the Emergency Operation Centre or the EOC. Each firefighter followed me through the front doors and received a day pass that was coloured blue with the word "Shovel" written on it. These day passes would change colour and have a different password on them to help keep only required people within the blockade.

The EOC was filled with a variety of emergency responders and support staff. Initially the goal of the EOC and the emergency responders was to help with the evacuation and rescue of people that lived in High River. Now that High River was in ghost town mode, the next objective was to make the town safe by pumping away the water that still covered over half of it and to remove the garbage and debris that saturated the area. The EOC had geographically divided the town into 10 sectors. In Sector 1, the northwest area of the town, the water that once flowed through the streets was now gone with only the flood residue left behind.

Groups of RCMP officers gathered in the parking lot after their shift. "You guys look hard-core," said one officer as we walked by his group. I nodded and looked back at my crew. They were covered in mud and most had Mora knives strapped to the radio harnesses on their chests.

It was shower and food time. The town was closed and the only showers available were in the school –but they were reserved for the RCMP. So we had to drive 40 minutes to a trucker stop to get cleaned up for the night. A food truck was available for us at the EOC and for the third straight meal, we ate hotdogs.

One time when I went back to the EOC I found Warhammer sitting beside the food truck happily eating a hotdog. Several paper wrappers were stacked beside him. Upon seeing me, he deep throated the wiener trying to hide the evidence.

"What are you doing?" I asked. He stared at me for several seconds trying to formulate an answer.

"What can I say? I'm a handsome man," he replied. Warhammer had been sneaking back to the food truck throughout the shift and had made friends with the hotdog lady who loved feeding the big man wieners.

It was strange driving around an empty town. But not all residents had left. Occasionally, someone that was hiding in their house would make a fleeting appearance outside or through their windows.

What was also strange was seeing boot marks on the front doors of a large majority of the houses in one area of town. At first glance it looked as if someone had kicked in each door with their muddy boots to loot the homes. But after inquiring about the boot marks, we found out that the doors had been kicked in by police looking for weapons. Residents were furious having their front doors boot fucked and then mud and sewage dragged in by the cops. It wasn't until someone with a brain told them to use a locksmith that the RCMP stopped treating private property like it was expendable.

On the morning of July 1, we went back to the downtown core and flagged off all the broken windows and other hazards in the area. We were told that the public would not be arriving until noon but they started to show up around 09:00.

We continued to help in the downtown area until we were convened to a meeting at the heavy equipment staging area. Our next objective was then made clear: work with the water trucks to spray the sidewalks and roads clear in each sector before the public could move in. And that's how the duration of the flood went for us. Find the next sector that was going to be open to the residents. Go identify the hazards and jobs that needed to be done. Wash the streets and roads. Then assist with the re-entry of residents.

For the next three days the sun was out in full force and the daily highs reached over 35°C. One day the temperature peaked at 38°C and for the crews working in the middle of the roads with no shade, it felt more like 45°C. The nozzleman could benefit from the coolness of the water but everyone else had to take intermittent breaks in the air-conditioned trucks to beat the heat.

Jordan and I went to Sector 3 and identified as many hazards as possible. With this information coupled with the hazard assessment maps that we had made for Sectors 1 and 2, we made a master map detailing all hazards in the sectors that we had worked in. We handed our information to the mapping specialist in the EOC and worked together creating a map of the town.

There were several more confirmed Warhammer sightings at the hotdog stand.

On our last full day on the flood, Sectors 1-6 and 9 were opened to the public. We had a couple of hours of work to do in Sector 5 before it was complete, and the public started to pour back in. We went to the fire hall in town for lunch. The burgers and fries were the first adequate meal we had been given since coming to High River.

We were soon assigned to clean up Sector 8 but the day before I had patrolled that sector and concluded that not a lot of work could be done until the water level had dropped. When we got to Sector 8, I was surprised by how far the water level had receded due to the "barge," a pump the size of a pickup that was pushing out well over three million litres of water a day. With the water gone we were able to wash the streets and prepare for another sector to be reopened.

After working 15 days straight, we had to time out and were released back to Manning. Over that time, we helped three communities. The devastation of the initial flood was behind us and the start of the rebuilding process for Southern Alberta could begin. We took three much needed days off before returning to base to see what the second half of the summer would have in store for us.

8

IN MEMORY OF A BROTHER

"My deepest thanks to my brothers. The lessons and sacrifice you made will never be forgotten. May I continue to carry on your legacy, though I know my time is short and borrowed."
– Brenden McDonough, Granite Mountain Hotshot lone survivor of the Yarnell Hill Fire

On freshly mowed grass in front of a Canadian flag that flew at half mast, I stood in front of a group of 70 or so people, 50 of those being firefighters in their best uniforms. The others dressed in suits and formal wear. A widow cried as her one-year-old giggled unaware of the ceremony. My good friend Brandon Taylor, a former Helitack leader, had borrowed my spare uniform to fit in with the other firefighters. Cavelle looked at me with kindness radiating from her green eyes.

"Hello everyone," I said. "It is good to see you all here. We have gathered to remember our friend and brother Matthew

Engelman." I took a deep breath to compose myself. "I first met Matt eight years ago at rookie camp and after that we were placed on the same crew. Later we would become leaders together for the first time. He was my friend."

My throat swelled with every word as I tried to hold back my pain. Eventually I was unable to speak and with the loss of my voice I was overwhelmed and wept freely in front of the crowd of men and women who had come to say goodbye to their friend. For the first time in front of my crew, I could not hold my composure and showing that side of me to the men and women I was leading felt wrong. I was supposed to be the strength of our crew and they saw me at my weakest. The raw emotion of losing my friend could not be contained, but hopefully seeing that side of me showed them the human aspect of how much you can care for the people on a crew.

There are events in life that take you by surprise and leave your world shattered. But none is more powerful than unexpectedly losing someone you love. This happened to me in March 2013 and no matter how strong of a person you think you are, grief will break you down and leave you broken. Matthew Engelman, my friend, died at the age of only 26. After failing his fitness test, he drove an hour and a half south from Peace River and took his own life under the Smoky River Bridge. Several years earlier, I had lost my older brother to suicide and losing Matt felt every bit as painful as losing Alan.

After our tour on the flood, we returned to our home base in Manning to gather in remembrance of our fellow firefighter who had died earlier that year. For the rookies, even though none of them knew Matt, this event was a solemn reminder not to take the time they had for granted – and to pay tribute to the man who had called Northern Alberta his home.

For the rest of us that knew him, this was a much needed personal experience, not only to honour our close friend who had fought fires in Peace River since 2005, but also a time for closure. As with most young people who die, Matt passed on without warning and left those who cared about him devastated. It was the first time that a fellow firefighter with which I had a strong bond with, died.

The memorial for Matt was held at the Manning Air Tanker Base where his family, including his wife and one-year-old daughter, and the firefighters gathered. Cavelle was flown out from her lookout to be with me. The last time I saw Matt, he had taken us both out for lunch, just days before he died. Cavelle helped comfort me during that loss more than she knows.

I had offered to give the eulogy for my friend. After I was able to speak again, I shared stories about our life together and the type of person that he was. When we first met, we were young and quite different from the men we had seen each other grow into. In 2008, he was my sub-leader on a Helitack crew where our bond was further strengthened as we learned how to lead together. We helped each other grow from young men into adults.

Matt's father passed away when he was young, and he often talked about how he thought his life would be cut short as well. Days before he died, he stayed at my apartment in Edmonton while passing through to start the fire season. I knew he was unhappy, but I did not see the extent of his sadness that he hid under his tough exterior.

Several other people took turns giving speeches or telling stories about their friend that they loved and would no longer be able to share their lives with. Ben Penner recounted a time where Matt's courage saved them when they stood their ground against an angry black bear.

After the ceremony, the firefighters who had been closest to Matt as well as his family drove an hour north to Twin Lakes. Matt was an avid fisherman and had enjoyed spending his evenings on the lake fly-fishing for the rainbow trout that were stocked there. That was the spot where his ashes would be spread.

As the sun lowered behind the white spruce that grew along the shoreline of the forest-green lake, Jordan strapped the urn that contained the ashes of our firefighting brother to his paddleboard. His long arms forced the paddleboard away from shore effortlessly with the paddle, each stroke moving him towards the centre of the small lake. As the sunlight hit the water, thousands of tiny bright reflections were illuminated in the slight rolling water created by the northern breeze. Every person on the shoreline shed tears as we watched the silhouette of Jordan open the urn and pour Matt's ashes into the water. When he landed back on shore, the tears that he had been holding back were now streaming down his long freckled cheeks.

I along with the two dozen others on shore stood around in grief and looked at one another without knowing what to do next or how to end the funeral that was taking place. Amid all the pain a voice rose in song that broke the silence, capturing the focus of everyone present. Ralph Cowie, a man in his 70s who had worked in forestry for over 50 years, had been a close friend to Matt. He sang a song from before my time. We listened with so much love from a man we all respected. A man who had devoted his life to forestry and wildfire. It was the greatest song that any of us had the pleasure of hearing. And it became the beginning of moving on that each of us dealt with in our own way, and in our own time.

Matt's funeral was a beautiful way to show respect for the man that he was, and I am hopeful that if he knew what we had done for him, he would have been happy that we cared as much as we did.

During that summer, a gazebo was built on the MATB under the leadership of Matt's good friend and fellow firefighter, Cole Pettifor. The green timber Gazebo with the red tin roof was placed by the pond where firefighters would relax after work and share their stories about the fires they fought, the people they loved, and the future they hoped to have one day. It stands in commemoration of our friend and reminds the Bomber Boys of Manning that even though Matt would no longer fight fires with us in person, he would still be with us in memory.

Depression and Emergency Services at times go hand in hand and without knowing how to deal with the stress or talk about it with your peers, it can become too much to handle. Whatever Matt was going through it became too much for him to deal with alone so instead of seeking help he decided to end his own life. He put an end to his pain but in doing so, he multiplied that pain by a thousand and spread it between the people that loved him.

It still hurts to think about what happened to my friend. At times I reminisce about the fires we had fought together and the friendship that I was blessed to share with him. Seeing him only days before he left us and talking to him on the phone the day before he committed suicide filled me with regret for years. How could I have not seen how much pain he was in? Maybe if I could have seen his pain, I could have helped him. Over time, I had to accept that what he did was his decision and that as a young man myself I did not have the skills that were required to help or to see that he was in that dark place where he felt he could not get out. Everybody grieves in their own way but in the end, acceptance, and forgiveness, not only for the person that hurt you, but forgiving of yourself, is the only way I believe you can truly move on. Sometimes life will give you the type of pain that changes you, but you must put in the work to let it change you for the better so inevitably, when hardships come again, you will be equipped to deal with that unavoidable heartbreak.

If you are in pain and feel there is nowhere to go, know that there is help. Suicide is final. I have had my own share of dark times and even though it feels like nothing can help, it can and will get better.

Matthew Engelman was a great person, father, brother, son, and firefighter. He will be remembered for his friendship, laugh, and willingness to help anyone that was in need.

He will always be missed.

9

SHITE CREEK

"People helping people."
– Kelly

Ten days after we left Southern Alberta we returned to the flood in High River. The first tour we had busted our asses. The hard work, the persistence, the pain. It had all paid off and we had been requested to come back to help once again.

By the time we arrived for the start of our second shift it had been a month since the first days of the flood. Most of the town was open to residents but closed to the public. The parts of town that were submerged when we were there last were free of water, but silt and debris still covered every surface that was once flooded.

After I checked us into the EOC we met with our strike team leader named Kelly, a fire ranger from the Whitecourt district. He was in his 40s with a heavier set frame and stubbled cheeks.

The EOC had a different energy than before. Even at this late hour, dozens of people were buzzing around inside the school, each reporting into separate rooms to whatever task they were assigned to, logistics, operations, etc. The Incident Command System that is utilized with Alberta Wildfire made us a perfect fit to help with this emergency. The experience Alberta Wildfire had with organizing a large work force to complete tasks was what was needed to clean up the last hurdles the flood had left behind before the rest of the town could start to rebuild.

As the leaders, Kelly, and I both stood in the main corridor of the EOC, he handed us maps that outlined the town and its current hazards.

"Oh sweet. We helped make these maps," said Jordan. They were replicas of those Jordan and I made with the mapping specialist on the last day of our last tour. Since then, the map had become the focal point for each meeting and a key in planning and prioritizing the workforce.

"I heard good things about your crew," Kelly said to me as we looked over the maps.

"Yeah buddy. We'll crush whatever you have for us," I replied.

"*Crush*?" Kelly said with a half smirk.

I chuckled at the look on his face. "It's an expression we've been using. Twenty firefighters can crush a ton of work," I said.

Kelly let out a boisterous laugh. "Oh, have I got a job for you guys." He pointed to the middle of the map. Our next assignment. The remediation of Little Bow Park.

We left the EOC and went to a new camp called Tongue Base set up in a farmer's field on the west side of High River. This camp was set up by a rig company and was luxury compared to what we were used to. We upgraded to ACTO trailers where each person got their own room, an excellent meal, and available showers – every night.

Early next morning we met Kelly at Little Bow Park. We parked by an empty shop beside Little Bow Creek which ran through the middle of the park. Immediately, everyone knew why this project had been left for a month. The entirety of the ground was covered in a thick layer of muddy silt that exuded a very distinct odour of human feces. The three-metre-wide creek which ran over one kilometre was lined with large aspen and poplar with an understory of thick willow which all of this caught the full brunt of the flood. The dense vegetation had captured every piece of garbage, refuse, detritus, flotsam, hell, whatever floated was sucked up into this natural web-like barrier. It was if every piece of trash in the town decided this was the place to meet up, get to know each other intimately, and multiply like a plague. Add to the fact that the flood occurred on garbage day, with all the garbage bins put outside the night before, well this didn't help the situation either.

Dozens of large garbage bins crowded the area around the parking lot, each toppled over with their rotten contents soaked in mud. Thousands of pieces of garbage and litter covered in foul smelling silt were visible everywhere. I don't know why I am calling it silt. It was more like shit but maybe that's what we had to tell ourselves to work in that godforsaken park. An odd-looking tool shed was lodged between two large aspens some one and a half metres above ground. The creek was clogged with garbage which stifled its flow and turned the water stagnant and foul. And all of this had been fermenting in the sun for the last past month, just waiting for a crew like ours to clean it all up.

The sight of the task set before me – or should I say us – was overwhelming. Low-key comments about how bullshit this task was echoed throughout the Unit.

"Have fun cleaning this up," Kelly said as we geared up for the massive job before us. "Garbage trucks will come by and pick up whatever you collect."

"Alright everyone. Let's just start one piece at a time," I told the crew. We were still building our reputation, and this was another opportunity to prove ourselves amongst our peers. Time to do the job that no one else was willing to do.

After equipping ourselves with rubber gloves and dust masks, everyone started to pick up the trash scattered around them. Piece by piece, a pile rapidly formed on one side of the parking lot. Echo crew was sent up and down the stream to complete any chainsaw work that needed to be done, including clearing danger trees or cutting free any woody debris that was holding onto garbage. Once again, they were asked to stay on dry land.

I tasked Alpha and Bravo crews to clean up the creek. Each firefighter going into thigh-high water was given a pair of hip waders to keep their uniforms from getting soiled. One by one they waded into the disgusting water, moving the garbage to the shoreline where the rest of the crew could collect it.

Garbage trucks showed up and we took turns loading up the filth where it would be shipped to the dump. Any chemical or biological waste such as car batteries or paint cans was separated into a different pile to be properly disposed of.

Garbage bags of all shapes and sizes were stuck in the mud or covered in heavy pieces of trash. Each time a bag was moved it would tear and release a rancid smell that would punch that unfortunate person in the face, making them gag. While Warhammer was retrieving a bag stuck in a tree above him, it tore, and a soiled baby's diaper spilled all over the front of his shirt, causing him to vomit. The rest of his crew gagged at the sight of the month-old fermented baby poo.

Swarms of mosquitos continuously attacked us as if they had a personal vendetta against the firefighters who were cleaning up their garbage haven that had seemingly been created just for them.

"Harold this is Marilyn. I need to talk to you," came a voice over the radio. I walked along a path cluttered in garbage that was only discernible from the absence of the trees and found her waiting for me.

"A lady came and talked to me a few minutes ago. Her daughter drowned and this was where her body was found. She was wondering if we would look for her daughter's purse that's missing," Marilyn said. It was the first time I heard that anyone had died from the flood and it took me by surprise.

"Of course," I replied. I broadcasted to the crew to look for any purses and set them aside for Marilyn to sort through. Quickly, dozens of purses began to show up at the parking lot, each one empty, full of mud, and with a price tag attached. A store downtown had been washed out and its contents had been caught by the willows that lined the creek. Sadly, that specific purse was never found.

Amid the cleanup, an unforeseen task presented itself. The high, fast-moving water had emptied many downtown-core shops, washing their contents downstream. James found that even though the task was gross, it was fun finding anything of value that had been washed away from its origin. While cleaning the creek he found a small back bag with a phone number attached and inside was a gold ring. He phoned the number and the woman who answered said that it was her fiancé's wedding ring that was getting resized. Within half an hour she made her way to where James was. Upon seeing her future husband's ring, she tried to hug James. He had to politely decline her embrace because he didn't want to get her clothes soiled by the mud he was covered in.

Over the events of the flood, this woman and her fiancé had become disillusioned and had questioned if they should go ahead with the wedding. They saw the return of the ring as a good omen and decided to go through with it. Later that summer, James would turn down an offer to attend their wedding. His spirits were lifted enough knowing he had helped the couple.

From the start of the season, James had mixed feelings about being on the Unit crew. He had done two years of Helitack, and his heart still belonged to his former program. He wasn't given the choice to join the Unit but because of his experience he was moved into a sub-leader role.

For James, it was special helping a community where he could immediately see meaningful results and he started to enjoy his new role. He found that when you became part of the Unit there was an energy in the air that you couldn't help but get sucked into. It surprised him how much work could be accomplished with 20 people working together and this flood reinforced that fact. What seemed like an insurmountable task set before the crew became real challenges we could overcome.

After getting the ring back, the future bride drove to the EOC to share what had happened. Her story spread throughout the EOC almost faster than the flood. The information officers that oversaw media relations loved it and made sure to spread the news about the goodwill of Alberta Wildfire. Others wanting to know about the origin of the ring found that it had come from a jewellery store downtown and that a substantial amount of that store's contents were missing. Once that information was known, I was instructed to look for any valuables that could have washed up in the park.

The crew worked diligently but it was hard to keep the morale high. It rained on and off throughout the day adding to the slipperiness of the muck. The rain washed the sewage off the leaves making the resultant precipitation something to avoid.

I had become proud of my crew and the efforts they put forward. This sense of family coupled with our accomplishments kept everyone happy throughout the season. But this day was different. Being covered in sewage and moving what only seemed a fraction of the garbage after a full 12-hour day had everyone feeling off. It was the lowest that I had seen the crew.

I knew that if this was the type of job they wanted us to do then we would need better protection from the garbage. Together with Kelly, we ordered protective suits, rubber boots, and proper respirators for the next day. We would not work in that foulness again without being fully protected. Someone more qualified than ourselves should have overseen this project making sure that we were properly protected.

Thank fuck we had our own rooms and showers after cleaning up that filth. If we had to stay in tents again, with gas station showers, and only hotdogs for grub, some of the firefighters on my crew would have broken.

The second day in the park involved the same objective: clean the park, pile the garbage, move the garbage into trucks. There were still massive amounts of garbage but the hope that was missing the previous day was now in the firefighter's hearts. With full-white protective Tyvek suits and better masks, the Unit was more than willing to get in the waste and clean that bastard park up. By the end of the second day, Little Bow Creek flowed clearly for the first time in a month. Paths hidden by mud-soaked garbage were visible and the source of the unwelcome odour that had permeated the downtown core was removed.

The Peace River Unit, in the words of the High River parks supervisor, had "transformed" the parks and walkways of the Little Bow area.

James, now called "the Raven" because of his proclivity for shiny objects, Courtney, and Couse, had found over $16,000 in jewellery and had returned it all to the EOC. With the returned booty, along with a returned wallet that Kelly found containing $1,000, and the remarkable cleanup we had done, the information officers and High River officials were loving us.

After Shite Creek, we cleaned the surrounding trails, fences, roads, and a dog park over the next few days.

Right from the start Kelly and I got along. We shared the same type of humour and seemed to appreciate each other's experience. It wasn't long before it was commonplace to see us laughing together over something that only we found funny. Over the years we would work together on several large fires, including a memorable tour in Idaho and Montana. It was always a pleasure to work with the man that could make me laugh all day even during the hardest of times.

After a week from our return to High River, many of the residents whose homes had not been destroyed moved back and the work needed from the Yellow Shirt Marines ended.

When the destruction was tallied, five people lost their lives from the flood and it caused over 1.7 billion in insurable damages. At the time, it was Canada's costliest natural disaster until the Fort McMurray Wildfire three years later. It was a rare time in Alberta Wildfire history where the men and women who battled flames were needed to help in another form of natural disaster. The art of hydrology coupled with the Incident Command System became a force that helped Alberta Wildfire fulfill a key role during the Southern Alberta Floods of 2013.

10

THE END WAS JUST THE BEGINNING

"They may not have been the best for fire, but fire was the best for them."
– Ryan Archibald, first Alberta Unit crew leader

It was the start of August when we returned from our second tour on the floods, traditionally the busiest time of year for wildfires. There were no large fires for us to fight and the season slowed to a crawl. The rains remained constant, and August 7 would be the last day that the Peace River Unit would be a 20-person crew. The start of university was approaching and the firefighters that doubled as students would have to leave for their studies.

Knowing it would be our last shift together, we went to the Peace Fest music festival and spent one last night celebrating our season partying to Sam Roberts. One by one, the firefighters left for their other lives and the Unit was broken up to take over initial attack roles or project work. Throughout the fall we fought several small fires but nothing we could not extinguish in less than a day.

In late fall, retired MATB camp boss and my friend Walt Tipton succumbed to cancer. Brando, Jordan, and I drove to his farm in Saskatchewan where we said our final goodbyes to the man that had watched us grow up and loved us like his own sons.

By October only Farley, Gary, and I remained, and on October 31, the fire season officially ended. The inaugural year of the Peace River Unit was over.

In total we fought nine fires, including five caused by lightning, two powerline fires, one farm fire, and one campaigner. Adding in the flood, the season was a success. We crushed every obstacle we faced and had set ourselves up well for the following wildfire season. From my experience working on incidents like these, I found it remarkable how much we accomplished especially compared to what I thought was initially possible.

When I reflect on that first season with the Peace River Unit crew, I am filled with emotions that bring a smile to my face. The expectation for success was set high but no one knew exactly what the formula for that success was when we started that season. With all the hard work, sleepless nights, and hours of our free time spent dedicated to the job, a foundation was set for future years. By showing our quality and what we could accomplish, Unit crews became a valuable and heavily requested resource on all future major incidents. After only the second year into the Unit crew program, everyone in Alberta Wildfire knew that when the best was needed, the Unit would be called.

Some of the best people that I have ever met were on that inaugural Unit crew and some are still my closest friends. James summed it up nicely when he told me how easy it was to get caught up in the energy that is the Unit. Comradery, family, and friendship were all extra benefits for each of these firefighters, even for those that only experienced that first season.

Almost every night after work that summer, especially on incidents, Jordan and I would discuss the day for hours, what worked well, what didn't, and how we could improve. Without the help of my good friend, that first year would not have run as tightly as it did. The unique way Jordan sees the world, his charisma, and his attitude that embraces the impossible, these are traits that I see in few people. We shared the same level of emotional intelligence that gave us the foresight to recognize that that time in our lives might never happen again – and we needed to soak up every drop of the experience.

For me as the leader on a new type of crew, everything did not always run as smoothly as I wanted. Every incident, training day, or project work came with its own learning points and tough times. Most of the crew were younger than 24 and for my 29-year-old self, being like a dad or older brother at times for the women and men on the crew was part of the job and another skill I had to develop. Looking after the crew's well-being helped me mature and learn aspects about myself that I had not yet realized.

There were times when I was hard on the people that I saw as hurdles we had to overcome. As the first Alberta Unit crew leader Ryan Archibald once put it to me, "Being salty and non-apologetic about the crew was what was needed in order for the program to survive."

The support I received from my supervisors Derek and Pat was enormous and it cannot be stated how integral their help was to the success of our crew.

Of course I am biased, but it was clear that the Peace River Unit stood out amongst our peers during the 2013 season. We fought the only newsworthy fire in Alberta and cleaned up the Southern Alberta floods. I received a personal letter of commendation from the premier of Alberta at the time thanking us for our hard work and the dedication the Peace River Unit displayed during the flood.

Working flood relief was a first for me and an experience that stands out in my 20-year wildfire career. Working with the public, especially those who were directly affected and were emotional, was another aspect of the job that was new to most of my crew. I was impressed with everyone on my crew and how professional they were even during the hardest of times. They all expressed how rewarding it was to help those people in need.

At each firefighter season-end review, I asked them what their favourite part of the season was. The consensus was they enjoyed their time working in Black Diamond above the rest. This was because of how much work we were able to do in such a short amount of time and how close they were able to work with the public. From start to finish there were always objectives to complete with no time wasted in helping the community get to the stage where they could start to rebuild. The town's gratefulness and appreciation were heartfelt by us all.

During times of fatigue and emergency situations, you get to know the people you are working with on a deeper level than other people in life. I am blessed that I had the chance to know all 19 firefighters on my crew in a way that most will never see.

I am grateful to everyone that I worked with that first season. I had so much fun with all of you. I am proud.

11

WINTER INTERMISSION

*"Being placed on the Peace Unit
was like being taken first round
by the Toronto Maple Leafs."*
– Michael "Soso" Sorenson

During the off-season, the firefighters of the Peace River Unit scattered to the wind, each leaving the Northern Boreal to seek refuge from its harsh winter. Only Gary and Farley stayed. They got jobs in Fire Smart, a project designed to help reduce the fire hazard by removing vegetation from lookout towers and other infrastructure in the district. Four months of stooping to pick up sticks five days a week left them unfulfilled, and they couldn't wait for the fire season to start up.

The firefighters who did not stick around went to school while others travelled the world or became ski bums on some resort in the Rocky Mountains.

As for me, I went to Australia to work as a wildland firefighter with the state of Victoria once again. I had worked there during the Black Saturday Bushfires in 2009 and wanted to go back Down Under where my experiences there had evolved me into the firefighter that I had become. During those fires in 2009, my crew and I got trapped in the town of Marysville as it burned to the ground. We all survived but 34 of the residents did not. I wanted to return to the place and be with the people that I had missed.

Cavelle came with me to visit the country she heard me talk about so often. She would not be returning to the isolation of lookout life and it gave us more time to spend together before the next fire season. Jordan even came for a two-week visit. With me turning 30 shortly after arrival, I would no longer be able to get an Australian working visa, so this would likely be my last time fighting fires overseas. It was a busy fire season that year. I fought 14 fires in the four months there. And altogether, I was in my 12th straight month of firefighting with another eight months ahead of me. I was madly in love, surrounded by good friends, and couldn't be happier with where I was in my life.

With time comes change and like fire seasons before, many firefighters who had given their blood, sweat, and tears for their crew did not return for one reason or another. Especially up North, the odds of a firefighter staying more than two seasons is about one in two, with the odds dropping exponentially every consecutive season. Maybe one in every 100 firefighters makes the job a career and stays in the wildfire service for at least a decade.

I enjoy asking each person on my crew if the job is what they thought it was going to be like before they became firefighters. Not a single person said it was how they imagined it to be. Most are surprised how physically demanding and time consuming it is to extinguish a wildfire.

The job, though, is amazing. You ask any firefighter past or present and they'll tell you it is the best job they have ever had. You can tell this by how many former firefighters still have a picture of themselves in uniform as their profile pictures on social media platforms.

The comradery, the action, the money, it's all hard to beat. But trumping all the positives is the personal life you must sacrifice to do the job. For four to eight months a year, you need to put your relationships on hold, be away from friends and family, give up city amenities, and live in the middle of nowhere. That is difficult for anyone and for those reasons among others, the job becomes a stepping-stone to other careers. The personal life you need to give up, as well as the high physical demands that wear and tear the body, keep all but the delusional from coming back year after year.

Over the course of that winter, five first-years and one subleader decided not to come back to the Peace River Unit for the 2014 season.

Sam "Dad" Hetherington transferred to the Helitack program in the Calgary District. Even though his heart was with the Peace River Unit, he transferred so he could be closer to his family and girlfriend who lived in Calgary. The nine-hour slog of a drive from Calgary to Manning is difficult for most and now his drive would be less than an hour.

Because Calgary is in the southern part of Alberta it is one of the slowest districts in the province with regards to wildfires. The five other lower districts also include Edson, Rocky Mountain House, Whitecourt, and Grande Prairie. These are not situated along the main lightning belt that passes through the province, so fires there are primarily caused by humans. The lack of fires in recent years created a huge turnover in the Calgary District and they needed experienced, mature leaders. Seeing that Dad had been to the two-week leader course, they offered him a leader position on one of their Helitack crews. After he phoned and listened to my advice that he was probably more qualified and ready to be a leader than anyone else in that district at the time, he accepted the huge jump in responsibility that up North would usually take several years to accomplish.

Marilyn transferred to the Rocky Mountain House District, another southern area, and joined a Helitack crew that would better suit her lifestyle.

Falling in love is one of the top job killers in wildland firefighting. This tragedy happened to Ian "Gadget" Megaga – and he was never seen or heard from again.

Colin McKinnon was torn about coming back for another season but the opportunity to teach English in Paris presented itself and it was too good to pass. He would return for another fire season in 2015 then get a full-time teaching position in Edmonton. Farley's father taught at the same school and Colin and Mr. Farley became good friends.

To the surprise of everyone, we discovered that Courtney was pregnant in the latter half of the 2013 fire season unofficially making the Peace Unit the first 21-person crew in Alberta. In spring she gave birth to a healthy baby boy and her new role as a mother would keep her from coming back.

Pursuing the life of a traditional family man, the Warhammer left the world of firefighting behind in search of the family he so yearned for. The sub-leader that created bewildered laughter from his actions remains a conundrum to the people he worked with. He is one of the few people in my life that I never truly figured out. I once told him that if I only had one friend in the world and it happened to be him, I would be blessed. Rumour has it that during the full moon on a hot summer's night, you can see the Warhammer's naked silhouette walking through the forests of Manning.

Naturally, Farley was promoted to the open sub-leader position and placed on Alpha crew so he could work closely with me. Over the course of the last fire season and two years of college, we had developed a strong friendship and I was happy to work and mentor my friend who was more like a younger brother to me.

There was only one transfer from the Helitack program to the Unit that my supervisor wanted: Desireé "Dezzie" Gerber. The Swiss-born farm girl who had immigrated to Alberta as a child had applied to be on the Unit the previous season. But due to her diminutive 5'2", 110-pound size, she was thought to be better suited as a Helitack firefighter. The four-person Helitack crews that routinely fly in light helicopters would ideally not take more than 1,000 pounds (450 kilograms) of cargo including firefighters and gear, so this weight restriction made her an ideal candidate. It was not her requested position, but she accepted the job anyway. She had originally applied for the 2012 fire season but was stopped in the recruiting process after failing the fitness test. With her stubbornness and unwillingness to quit, she recreated the fitness test by building a ramp in her barn and replicating the weighted pump and hose bag. She practised almost every day for a year and passed the fitness test the following spring.

During the end of the 2013 fire season, I filled in as a Helitack leader and got to work with Dezzie on a couple of fires. I was impressed with her work ethic, especially when she helped pick up the slack of the other firefighter we were stuck with, and I put in a good word for her transfer.

Early into the season, starry-eyed Matt Dawes accepted a job as a structural firefighter for the City of Fort St. John. Gary was heartbroken. Since the season had already started there were no more rookies to pull from training. This meant that the Peace River Unit would be a 19-person crew for the year unless they could find a previously trained firefighter. I knew the perfect person to fill the position: Brandon "Brando" Taylor.

Brando was a trim 6'3" man from Edmonton who enjoyed fishing and rock climbing in his spare time. He started Helitack in Peace River in 2004 and worked there for eight seasons during which time he got a diploma in Forest Technology from NAIT as well as two degrees from the University of Alberta: one in Forestry and the other in Reclamation.

During the seven years of working Helitack together, Brando and I developed a strong friendship because of our common passion towards fire, fishing, and the outdoors. We continued to be friends after Brando decided to move away from wildfire to put his schooling to use and pursue a career in the oil and gas industry by reclaiming well sites and tailing ponds at Fort McMurray. After a year, reclamation work became increasingly scarce, a foreshadowing that the Alberta oil and gas boom was ending. The prices for the easily accessible black gold that had propelled Alberta into an economic powerhouse were failing. Who would have thought that demand for high priced oil would last forever? Alberta did. With China's dependence on oil dropping and the province not being able to compete against cheap oil drilled out of Saudi Arabia, the great Alberta oil boom of the early 21st century was over. Tens of thousands of jobs disappeared into thin air and a recession crept its way into a province that was once so rich that it's late Premier Ralph Klein gave every Albertan man, woman, and child $400 – just because there were so many billions of surplus dollars in provincial government coffers.

Luckily for Brando he could fall back on his wildfire experience. There are always going to be fires, which creates a certain amount of job security to those who are willing to fight them. Even though he wasn't pumped to be a red hat again, the promise of guaranteed work and the fact that he would not have to answer to anyone except me was enough. Brando still felt the loss of Matthew and coming back for another wildfire season would give him new memories to help cope with the grief that he still held in his heart. With him came an added wealth of firefighting knowledge and forestry experience that could not be found anywhere else in the province.

In its sophomore year, the Peace River Unit was now a powerhouse compared to its previous self. Its combined experience had grown from 24 seasons at its conception in the spring of 2013 to over 45 seasons. The foundation had been laid and high expectations set. Even though the program was still in its infancy, we felt ready to take on any test that Zeus or Prometheus could throw at us. And tested we would be in ways that would push us to our physical and mental limits, bonding the crew of 2014 like no other.

The 2013 Peace River Unit Crew
Top Row: Harold Larson, Warren Keeler, Jordan Sykes, David King, James Williams, Mark Alexandrino

Middle Row: Matt Dawes, Ian Magega, Nate Thompson, Colin Mackinnon, Sam Hetherington, Courtenay Ferguson, Gary Thirnbeck

Bottom Row: Mike Sorensen, Jesse Hoevenaars, James Couse, Adam Fulthorpe, Andrew Farley, Andrew Fulthorpe, Marilyn Morand

Fire jumping the Forestry Trunk Road on the Fish Lake fire.

Project work at one of the many lookout towers in the district.

A Unit firefighter balancing on the skids during hover exit training.

Lightning south of the Manning Air Tanker Base.

Matt Dawes sandbagging the Elbow River.

Aerial view of High River during the 2013 Southern Alberta floods taken by Jordan Sykes.

Matthew Engelman 1986 – 2013

The 2014 Peace River Unit Cre

Top Row: Julie Gillis, Colin Popp, Mike Sorenson, Nigel Thompson, Quentin Schmidt, Joel Pecotich, Adam Fulthorpe

Middle Row: Desiree Gerber, James Couse, Nate Thompson, Jesse Hoevenaars, Matt Dawes, Andrew Fulthorpe

Bottom Row: Mark Alexandrino, James Williams, Jordan Sykes, David King, Harold Larson, Andrew Farley

Brandon Taylor with a monster northern pike out of Sawn Lake.

The Peace River Unit conducting a hazard reduction burn in the Peace River valley.

Nate, Andrew, Jesse, DK, Julie, and Couse on a loaded patrol searching for fires over the Clear Hills.

Alpha and Delta crews getting picked up after crushing fires in the High Level district.

Bravo Crew on the riverbed of Fire 43.

Bravo, Charlie, and I waiting to be flown back to base after watching the airshow.

An 802 unloading retardant on the north excursion.

Air tanker dropping on the ridgeline of Fire 43.

The natural barrier of the south rock face getting tested.

The south rock face the next day.

Jordan Sykes calling in a helibucket to fill a bladder on pride rock.

Popp, James, Jesse, and I at the 2014 Alberta Wildfire Olympics.

12

SPRING TRANQUILITY

"Best job I've ever had."
– Andrew Farley

The fire season started once again on March 1 when the North was covered in winter. The mighty Peace River flowed relentlessly under a thick sheet of ice flanked by snow-covered farm fields and forested hillsides. The time before the great melt would be used to get ready for the upcoming wildfire season. While the fires had yet to burn, it gave us the time to train and certify the crew on ATV, chainsaw falling, and helicopter hover exit.

Until the ground thawed enough to use underground water pipes, the Manning Air Tanker base would remain closed, and we would instead live in the Sawridge hotel in Peace River.

In between training, we provided fire permits to farmers of the North. These permits allowed them to continue the practice of burning off old crops or infrastructure if they stayed within guidelines. If they burned outside of legislation, then we would issue an Order to Remove the fire hazard, which depending on what was burning and their available resources, they would have a certain number of days to extinguish their fires.

I enjoyed this part of the early season as it gave me a unique view into the lives of the people who lived there. Otherwise, I had no reason to go to the massive Mennonite farms that expanded every year at a rate that I had a hard time comprehending. These old-school Dutch colonies lived in complexes that housed dozens of families where most of the children looked eerily similar. They embraced technology but continued to live a traditional religious lifestyle. Each year, these hard-working farmers consumed vast areas of the northern boreal forest, transforming it into farmland.

The time of single families running farms seemed to be a part of the past. Farming companies or the Mennonite colonies own most of the farming land that lined the main roads in the district.

Most of the fire permits were given but sometimes they were denied for safety or environmental concerns. A common belief among farmers is that putting out a burning slash pile or stopping a wildfire was easy and they could handle it themselves.

On one occasion, Farley and I went to an elderly couple's farm where they wanted to burn a large pile of trees and dirt less than 10 metres from their house.

"Well, that pile is dangerously close to your house and with the resources you have, I can't allow you to burn that," I explained to the 90-year-old man who stood in his doorway. On cue, his wife's head popped around the doorframe.

"This is bullshit," she creaked from behind her husband while he blocked her inside their home.

"Martha, that's enough, we've talked about this," he calmly counselled.

"Alright, I guess. Let us know if you move that pile and give us a call," Farley said not wanting any more outbursts from Martha. Farley and I could not stop laughing about how the lady had appeared out of nowhere to express her feelings and her quote became a running joke for the two of us.

For most of April, we were assigned to help prep the Hutton prescribed burn in the Peace River Valley. This 500-hectare area was originally burned in 2007 was slated to be re-burned. The intent was to burn off the remaining old growth of trembling aspen and brush along the river to create ideal elk habitat. By opening the area, the hope was that the elk would breed and stay in the area. I had worked on the original burn and it was great to see it completed. By May 8, the burn was finalized with only two excursions on the main burn needed that day which were both easily contained by us. Once again, working on a prescribed burn gave the rookies first-hand fire experience and time to work with the tools of the trade. Establishing that clear chain of command and giving the rookies opportunities to prove themselves before stepping on an out-of-control wildfire – that would be key for our success.

By May the ground had thawed enough for underwater pipes to function in Manning so the MATB opened. At the hotel in Peace River before that, the staff changed the white towels for green for the firefighters who had a hard time cleaning themselves properly; after a hard day's work they were leaving the white towels covered in dirt and ash. Hotel life is fun but after two months it was a nice change to get back to my hidden paradise of Manning.

By May 12, the Peace Unit was a complete 20-person crew after the last of the firefighters arrived from university. Through the rest of the month, we helped the Manning Volunteer Fire Department with hazard reduction burns in the fields south of the base. We burned off dead fields and ditches along the highway, reducing the fire hazard in the cured grass that had accumulated over winter. By that time, some of the sophomore firefighters had seen more fire than they had all last season. In fact, we had already spent twice as many days working with fire compared to our previous season.

At the end of spring, we were base-changed from the MATB to the Chinchaga Base. This remote camp contained two sleeper trailers, a cook shack, and an office located at the end of a gravel road 80 kilometres northwest of the MATB. I was told by a former wildfire supervisor, Lookout Ralph Cowie, that the Chinchaga Base was once the location of a minimum-security prison. The beautiful camp had been built in an oxbow of the meandering Notikewin River and would have been a pretty wild place to serve out your time. For some of the city people on the crew it probably felt like they were doing time. No cell service or contact from the outside world – just the mosquitos and bears to keep you company.

But if you looked closely, it became a place of wonder. Belted kingfishers dove off branches of white spruce into the river, spearing fish with remarkable accuracy. The hermit thrush's famous song whistled through the forest. Small Arctic grayling with their oversized dorsal fins searched the river for insects. Every summer I enjoyed coming back to this camp to retreat into the wilderness. After each workday, Brando, Jordan, Marco, and I would go to a firepit on the outskirts of camp that we had built along the riverbank so we could appreciate the comforts of the outdoors.

It was at the Chinchaga base on the last day of May that we fought our first wildfire of the season. Storms had passed though, and several fires had starting from lightning. With a helicopter, Alpha crew, Bravo crew, and I flew to a smoke that had been called in which was close to the Doig Lookout area. The fire burning in the middle of the boreal forest was contained to 1.7 hectares by several tanker drops, a size that was manageable for the eight of us. The Bird Dog that was in control of the airspace would not clear us to land in the opening beside the fire. Not wanting to wait for the Bird Dog, we found a well site where we landed and hiked the 3.4 kilometres carrying over 250 kilograms (550 pounds) of gear to the fire. The hike was a struggle for some of the firefighters. Carrying that much gear through overgrown seismic lines where with every step your boot sinks into the mossy ground, filling it with water, compounded the task before us. On arrival, the fire intensity had diminished to a creeping ground fire. Water was readily available out of sump holes and the fire required several hours of work before we could call it extinguished.

The following day, I, along with Charlie and Delta crew, flew to a 0.25 hectare fire just inside the district of High Level. We hover exited into a waist-deep pond on the edge of the wildfire and along with two sub-crews from the Grande Prairie Unit, we were able to extinguish it in less than two hours.

Otherwise, the hazard remained low and the province remained idle of wildfires, so for the second year in a row, Hinton hosted the Alberta Wildfire Olympics. Nineteen teams representing crews from all over the province again competed to see who would be crowned Alberta's champions.

Last season, our team was represented by Farley, DK, Dad, and Soso. They finished midfield in 8th place, but I knew we could do better. Last year I could not compete because I didn't want to leave my crew if a fire happened. Now that I had trusted leaders, I took it upon myself to represent my district and show Alberta what my crew could do. James, Jesse, and Popp joined the cause to see how we stacked up against the competition.

Over three days, 10 events tested our skills and fitness at a gruelling, competitive rate that left us all exhausted. Half the events were done in the pouring rain, including a campfire-building event that after hours in the downpour, only five teams were able to complete, including us. Other events included GPS and mapping, shelter building, a hose relay up the steepest hill in Southern Alberta, multiple fitness tests, a cook-off, and wilderness survival. At one point during the gear climb, I blacked out after leaving everything I had in me on the field.

We placed high in all events and got a silver medal in the cook-off and tug-o-war, which was by far the hardest competition there. At the end, we fell a couple points short and placed 2nd overall in an extraordinarily strong field. It was the hardest competition that I had ever taken part in. If only we had homecourt advantage …

We finished our 15-day shift in Chinchaga and when we returned from days off, we were base-changed to Haig Lake, 200 kilometres southeast of Manning. Our new base was located by two of the local First Nation reserves: Cadotte and Little Buffalo. Less than a kilometre from the base is Haig Lake. This large but shallow lake is a hidden gem in the North. It is a favoured spot for the crew to fish for the large pike and walleye that lurk in the dark green waters.

June 21 was the summer solstice and by 11:30 pm the sun barely sunk below the horizon only to rise again shortly after 4:00 am, making the long days perfect for the opportunistic fishermen on the Unit.

That evening after work, Jordan, Brando, and I went to the lake with Jordan's inflatable zodiac, the *OMC Express*, equipped with a four-horsepower motor that we had stowed in the trailer that held our gear. Within minutes of launching from shore, the fish were hitting. Something in the air or water or atmospheric pressure was making the lake come alive. Our lures, white jigs with pink heads and red devil spoons, would just hit the water before a fish with sharp teeth and pitch-black eyes attacked it.

Large white pelicans that prove dinosaurs once ruled the earth soared mere feet above water looking for fish to feed on.

Around midnight, the bluebird sky faded into twilight causing dark hues of orange and purple to reflect off the water. The haunting calls of loons echoed as the sun set behind the forest that enclosed the lake. But even with the sun hiding behind the trees, it remained light enough to see the entirety of the shoreline. At times the three of us would simultaneously have a fish on the line causing joyous pandemonium on the *OMC Express*.

Cheers could be heard from Farley and Joel who were sharing in the same success from a canoe. Joel was one of the five rookies that had come to work from overseas. While Jordan was travelling through Sydney, he met Joel while working as a city arborist. They developed a quick bond and Jordan suggested that he apply for the Peace Unit. This fun-loving Australian with infectious laughter was a bright light on the crew and made friends wherever he went. Like Jordan, his free-spiritedness could sometimes get the better of him as he pushed the limits at work.

We laughed, told stories, reminisced, and shared our thoughts. We missed our friend Matthew and wished he could have been there with us. By 2:30 am, when the fish calmed and stopped assaulting anything that moved, we had caught and released 44 walleye and northern pike.

"This will be one of the greatest memories that we will ever have," I said on the boat.

"What do you mean?" Jordan asked.

"Man, in 30 years I have never had a fishing trip like this and to share it with my buds on a night like this – it doesn't get any better."

I knew that I could not take what I was experiencing for granted. This was going to be my last summer leading the Unit and working alongside my friends. I had a moment of perfect awareness of exactly where I was in the world. My soul was content.

Jordan and Brando remained silent and let that possibility sink in. "Yeah buddy, you might be right," Jordan said. That would remain the best fishing event of my life. That night became a catalyst for our friendship that bonded us in a unique way. Every year, no matter how far apart we live, the three of us meet up for a fishing trip somewhere in Alberta.

It was the inner calm I needed before the second half of the season. It was midsummer, and we were now in our full routine. The season was steady with meaningful work that had given the Unit experiences to build upon. The work ethic instilled in my crew would help establish the baseline for what I wanted to see in them daily. The work ethic that was needed for the fires to come.

13

THE FIRE FLAP – PART 1

"Someone poisoned me."
– Colin Popp

"Do you want to take my spot today?" Popp asked Julie. The Bravo and Charlie crews were assigned to be in the helicopter for initial attack duties. Since they would be joined by me and Gary, the seats were limited, and one of the two rookies would have to stay behind.

"Aye, but are you sure?" she responded.

"Hell no, but we might get another fire today and I have more than you do," Popp said as he ran his fingers over the two long patches of hair growing under his chin. It had been a month and a half since he shaved and he was annoyed with his inability to grow facial hair.

As an initiation to the Peace River Unit, each new firefighter was asked to not shave until all the rookies were "certified firefighters," meaning that they had fought five fires or spent 150 hours on the fireline. Once a firefighter is certified they can be exported, meaning they are allowed to fight fires out of Alberta. This is a very cool experience that always generates a lot of overtime. But once fires are burning and a province needs firefighters, no one gives a shit if you're certified, and they would be exported anyways. It was a fun initiation that experienced firefighters loved to share with their rookies.

Since we arrived at Haig Lake the week before, we had fought two fires caused by lightning: a small 0.1 hectare muskeg fire and a more challenging 1.5 hectare fire that the entire Unit had fought and crushed in a day and a half. Popp, a 22-year-old muscular Calgarian, had fought both those fires and wanted to share the fire numbers with his friend.

Julie, a 27-year-old and the oldest of the five rookies, was a recent graduate from Xavier University on the East Coast. Coming from a family of 10 siblings who were mostly male, the woman with pitch-black hair and a thick Gaelic accent fit right in amongst the Bomber Boys.

Popp had fought three fires, two more than Julie, and was close to being certified. His selflessness would later be a decision he would regret.

It was the morning of June 28 and I was getting the crew ready for the morning briefing. Standing on the office deck, I scanned the sky to the north over the short grassy field that tapered off into a thick stand of trembling aspen. Cumulus clouds were scattered across the sky. The clouds resembled soft cauliflower heads, indicating that the potential for lighting was imminent. I had seen that type of atmospheric instability hundreds of times before and this was no different. The humid air was as thick as the anticipation of what the day would have in store for my crew.

Rain had been scarce over the last week which added to the probability that fighting a fire was in our near future. While briefing the crew, dried grass crunched under their steel-toed boots as they gathered to hear about our objectives. Most of the firefighters had moustaches that they had cultivated over the winter. I too had grown a thick moustache that had become part of my wildfire persona.

We would have a Bell 212 helicopter that could fit eight people. It was challenging to configure the crews for the upcoming tasks. Each of the firefighter's skills as well as how their sub-crews worked as a team were considerations I had to base my decisions on. I tried my best to make it fair but sometimes you need the right people for the job. Knowing that the day was going to be a banger, I wanted Brando to be there with me. Not only was he one of the best firefighters on the crew, but he had also just come from transitioning careers, and I wanted to remind him just how amazing this job was.

That put Bravo crew of Marco, Brando, and Dezzie in the ship with me. DK had a wedding that he had to attend so I took the three members of his crew, Jesse, Julie, and Couse. I wanted some more experience on the crew, so I took Gary from Delta. Since we could only take eight, Popp made the call to give up his seat putting him on Echo for the day.

The eight of us would be on five-minute getaway, meaning we had to be up in the air within five minutes of getting dispatched to a fire. With the pilot needing at least three minutes to spool up the rotor blades and get the machine off the ground, our only option was to wait by the helicopter or "sit on the skids" to make our getaway time.

My plan was to have Marco sit in the front seat of the helicopter and be the conduit between our crew and dispatch. I had been in that position thousands of times and I was able to mentor new leaders. I enjoyed teaching and sharing my knowledge. It gave me a different sort of satisfaction knowing that I was passing on what I had learned throughout my years while I was growing as a leader.

The rest of the crew would be ready at base in case we needed the support of another crew. If a fire was more than the eight of us could handle, I could send the helicopter back to pick up the rest of the crew unless bucketing on the fire was the priority. The crew could also be dispatched to other fires if the need arose. My leaders had earned my trust to fight fires without me and to keep the crew safe.

My work cell vibrated from inside the right chest pocket of my ash-stained yellow Nomex shirt. Grabbing my phone with my left hand I simultaneously reached for my notepad and pen from the right cargo pocket inside my forest-green Nomex pants. Seeing "unknown" on the call display I knew that it could only be the duty officer. I pointed my index finger in the air and made a circling motion. In unison, the seven firefighters scrambled from their seated positions and rushed towards our helicopter.

"Harold here," I answered with my standard greeting that I had been using at work for over a decade.

"Hi, Harold. Just a patrol," said the deputy duty officer. I raised my free hand creating a chopping motion at my neck signalling to the crew that the call was not for a fire. It was common to get at least one patrol a day to fly around and look for fires and the crew relaxed knowing what the call was about. Most of the time in my district, when a crew has a helicopter for the day, they would be tasked with a loaded patrol[6] and fly in a predetermined route looking for smokes that may be hidden from the lookout towers. The 5.2-million-hectare landscape that covered our district was almost the same size as Nova Scotia.

"No worries. Where are we off to today?" I replied. The duty officer relayed several coordinates and lookout towers to fly to that met the coverage standards a computer program had calculated for him.

"Right on. We'll be airborne shortly."

I opened the district map on one of the many apps on my cell. Looking for given townships and ranges using the "Alberta Third System of Survey" that delineates the province into 1-by-1 mile grids, it was indeed going to be a typically long patrol.

I love flying in helicopters and how it gave me a unique look at my district from the air. Even then, I was not stoked that the patrol was going to be at least an hour and a half. The comfort of passengers in the back seat of these things seemed to be an afterthought. The design of two metal bars that held a thin canvas sheet prioritized placement of large cargo, not our backs and butts. That combined with the constant "whoop whoop whoop" that can

---

6   Loaded patrol included our gear: two chainsaws and chainsaw kits, four boxes of hose each 60 pounds (27 kilograms) and containing 400 feet of hose, a Mark III pump and pump kit, jerry can of mixed gas for the pump, 2 smaller "combie" cans of fuel and oil for the chainsaws, four Wajax bags, backpack full of drinking water, eight hand tools, and personal backpacks. Each personal backpack contained spare batteries for our radios, Stanfield sweaters, raingear, food for 24 hours, and various other tools or amenities that suited each person. Each leader had a GPS, satellite phone, and a weather-measuring instrument. Depending on the make of the helicopter, our total weight including the crew usually hovered between 2,000 to 2,400 pounds (900 to 1090 kilograms). The gear is stored in the sides or "hellholes" and the tail boom of the helicopter.

subtly cause you to fall asleep – all this was not particularly enjoyable after the excitement of flying wore off.

The crew loaded into the helicopter, taking those positions they had picked after briefing. I took the forward-facing seat behind Marco while Brando took the spot by the opposite window so we could have the crew's best vantage points on each side of the helicopter. The rest of the crew filled in, as usual with the newer members in the middle facing backwards.

Marco climbed into the front seat, skillfully maneuvering his legs around the collective lever and steering column. The collective would be positioned to his left by the door and the steering column would be between his legs. Most helicopters have dual controls so they can be piloted from the other side while bucketing as the bubble window provided a better view of the world below. Strapping my black kneeboard to my leg where I could write notes, I buckled my seatbelt, put on my headset, and waited for the helicopter to start its lift.

The pilot began his checks and started up. Flipping several switches to initiate fuel and other key components, the whirlybird was ready to fly.

"All good back here," I said over the headset, indicating that the crew in the back was buckled up and ready to go.

"Sweet. Let's do this," said the pilot.

His lips were moving as he spoke into his mouthpiece but he could not be heard over the headsets; instead he was now talking on the air-to-air channel making sure the airspace was clear. Moments later, the helicopter lifted into the air. The ground became smaller as the horizon grew larger. Every time we took off into the sky, I was amazed at how awesome my job was and how grateful I was to be in that position. And this time was no different to the thousands of other times. Now muscle memory kicked in and I noted the movements of our patrol.

Pilots are a different breed of human. Crews would come and go but pilots seemed to remain, so at times I came to know my pilots better than my crew. Building trust with pilots allows you to put your life and the lives of your crew in their hands. The culture that is created between pilots and firefighters of the North is special. Pilots love their profession. As the old joke goes, "How do you know if there is a pilot at a party? They will let you know."

I could always tell the level of skill a pilot possessed by their pre-flight briefings. Every time we used a new helicopter or got a new pilot a briefing is required. After hearing hundreds of briefings, these two observations became truths: 1) The more direct and usually funny the briefing was, the higher the skill of the pilot and, 2) The long-winded pilots who spoke to us like shaved apes about to step into a magical machine were usually trash.

Our pilot Shane was charismatic, humorous, and his briefings were flawless. His charm when it came to the ladies combined with his taste for adventure made for some interesting stories on our patrols. I had been working with him for over a decade and he was one of the best in the business. On more than one occasion, Shane referred to my crew as professionals, sharing our mutual respect.

Being on patrol, flying over the countryside searching for anything out of the ordinary, is a wonderful experience. The relative flat landscape extends the view for hundreds of kilometres. Around that eastern area of the district, the landscape at first seems unremarkable but with each patrol, the beauty of the northern boreal is revealed. Gentle rolling hills covered in a mix of conifer and deciduous trees, pocketed with pools of water each having a beaver lodge built in them. Decades old seismic lines cut through the forest in every direction creating a man-made presence throughout the forest mosaic.

When I first started fighting fires in the Peace River District, it was common for us to land close to a seismic line and use the cleared-out three-metre-wide paths to reach our destination. Over time, the forest had naturally reclaimed itself and these once open pathways were so overgrown with willows and alders, it was once again easier to walk through the forest than through the tangled mix of vegetation that was competing for sunlight.

Close to the Hawkhills Lookout there is an area that is absolutely unique. For 10 minutes straight, we flew over thousands of islands covered in trembling aspen and balsam poplars each surrounded by dark shallow water. Each body of water no bigger than a basketball court had a beaver lodge in it that must have been home to thousands of those large furry creatures. The bright white plumage of trumpeter swans stood out amongst the dark background. These rare, majestic birds are always found in pairs, never more or less. From high enough in the air, the patchwork of green leafed trees and brown water resembled a cross section of a lung, with each body of water connected like that of a bronchial tree.

The mosaic of water bodies and forest transitioned into farm fields of various shades of brown and green that surrounded the farming communities of the Hawkhills. The bright yellow flowers of the canola fields stood out amongst the other crops. Each field was broken up into quarter hectare squares that were once a standing forest that had long since been cleared for agriculture. Tractors were running everywhere, some creating a column of dust that rose behind them as they tilled or cleared their fields. Dust columns in the distance sometimes would appear as smoke causing the newer firefighters in the back of the helicopter to ask if it was smoke. The subtle difference in the brown hue of dust separated itself from the bluish-grey tints that were produced with smoke – experienced firefighters would let rookies know this difference.

At the halfway point of our patrol, we reached the farthest northeastern lookout in our district, The Hawkhills. This lookout housed Don Ross, a unique man that had been doing the job from before my time in wildfire. I am not sure if the solitude made him the way he was, but eccentric is the only way to describe him. He would walk around with several bells attached to his body to warn off bears, label everything in his lookout cabin, and describe in detail how magnetism is the next energy that we need to harness. His weathered black kneepads that he wore over his faded jeans to keep his knees from getting banged up while climbing the tower added to his look.

Nevertheless, each lookout reminded me of Cavelle and how far away I was from the other people that I cared about. As much as I loved the job, I couldn't be gone forever. But now wasn't the time to think about home.

The patrol lasted, as expected, an hour and a half. By the time we began our decent to the Haig base, the towering clouds in the distance had doubled in size since morning. Their colour had transitioned from white to dark grey. The first crack of lightning highlighted the ominous storms to the north.

"Here it comes," I said over the intercom. "The show's about to begin."

14

THE FIRE FLAP – PART 2

"They breed a different kind of firefighter up there."
– Calgary Operations Chief, 2016

The dark clouds continued to build upon themselves. Each particle in the storm created a static charge and when enough electricity built up, the force was unleashed. To the north, loud cracks echoed from lightning followed by the deep rumble of thunder.

An unknown caller flashed on my cell. "Get in the sky and head north. Multiple starts. You'll get coordinates in the air." It was a dispatcher.

"Copy that," I replied.

The seven other firefighters scrambled to their seats in our helicopter before the call was finished. Shane started the engine. The rotor blades grew louder as they picked up speed.

At 13:47, we lifted from Haig Base and flew towards the dark clouds. Multiple storms could be seen at once, each throwing lightning ahead of its path. Rain followed from the hulk of the pillowy giants.

"Got visual on the smokes," Shane said. "Looks like a couple of them at our 11 and another at our three." Pilots always seemed to find fires before I could spot them. Focusing my vision, three light-grey smokes revealed themselves.

Farther in the distance, several other smokes rose on the horizon. Sunlight reflecting off the Bird Dog gave away its position as it circled those smokes. Our helicopter's radio buzzed with traffic. The Bird Dog air-attack officer replayed to dispatch that he was close to four smokes. Each fire was less than a hectare in size but with the incoming storm, he would have to delay the air tankers until it was safe for them to fly in that area. The AAO confirmed those smokes were north of the invisible line that separated the Peace River and High Level districts. For now, those fires would be High Level's problem.

We approached our three smokes. The closest turned to flames as a dozen black spruce trees torched, shooting flames into the sky. As rapidly as it ignited, the flames died leaving smoke coming from the charred trees. Dark clouds enveloped the skies overhead and rain poured onto the smokes, hiding them from view.

To the west another storm raged, spitting out lightning at quick intervals. With the rain, lightning, and downdrafts coming from the dark clouds it was unwise to fly into that storm. In the past, I have pushed my pilots closer to storms than was necessary, resulting in some close calls. Years before, my helicopter had been so close to getting struck by lightning that I could feel the electric shock through the mouthpiece of my headset. It left my lips tingling. Not wanting to gamble with the lives of my crew, we would have to delay landing.

"XMA Two-Six this is Kilo November Juliet," Marco relayed over the radio.

"Go for Two-Six," the dispatcher replied.

"You can check we can see three separate small smokes that are not growing. No values at risk but we will have to wait for the storm to pass before we can land."

"That's all copied. Two-Six."

The storm was pushed by strong winds from the passing cold front and was tracking quickly to the east. Our pilot slowly turned the helicopter to the west as we began a wide circle looking for any other fires that might pop up. After flying around for 15 minutes and not finding any other starts, we were comfortable with the distance from the storm and went to look for the three fires.

Easily enough we found the southernmost fire that had been contained to a handful of blackened trees. The rain that escorted the lightning had kept the fires from growing since our initial assessment. A fire that small didn't warrant the attention of eight firefighters. That meant we could split up and tackle two fires at once.

Marco confirmed the wildfire with dispatch and filled the "FP48" or "white" message that would paint a picture of the fire behaviour, request other needed resources, and potential firefighting difficulties. After dispatch responded that this wildfire would be numbered "PWF 056," Marco relayed the 16 pieces of required information to dispatch then let them know that we were going to land close to the fire.

From our perspective high above ground, the hike to the fire looked close to a clearing that we could land in. But with elevation comes distortion. Even with the distance to the fire not much more than a kilometre, we didn't take for granted how taxing the hike in the muskeg would be. After a few circles, the helicopter landed, and the four firefighters of Bravo crew exited the helicopter. They unloaded a chainsaw, Wajax bags, rubber backpacks that could be filled with water and sprayed from hand pumps like a water gun, and their personal bags – all the while the rotor blades continued to spin above them.

I got out of the back seat and instantly my pants were soaked from the moisture attached to the tall grass. The downdraft of the blades caused the ever predominate moisture to seep into our Nomex. It was time to embrace the suck. There was no way to avoid getting saturated from head to toe and those that try to remain dry fail and become miserable. The time to be warm and dry would have to be postponed until the end of the day that was still over 10 hours away. I waited for Marco to carefully get out of the front seat and once he was clear from the door, I patted him on the back and leaned close to him.

"See you in a bit, brother," I yelled over the roar of the rotor blades. He nodded with a smile. I transferred to the front seat, being diligent to gracefully lift my leg over the dual controls. Once I was strapped into the four-piece harness seatbelt, Marco gave the thumbs up to Shane signalling that his crew was ready and that all doors and compartments of the helicopter were closed. Bravo crew huddled together beside the machine as the helicopter lifted off the ground. The rotor wash became powerful causing the crew to tightly hold their gear not wanting anything to blow away or get sucked into the blades. It's a conundrum how strong yet fragile a helicopter can be. I've seen the smallest damage ground a helicopter for days, yet I've also been in a helicopter that had its blades strike small trees only to have the pilot shrug it off. As the helicopter skids left the grassy field, the burst of power from the rotor wash caused the crew to lean forward, negating the force that was on the verge of pushing them over.

From my vantage point the four firefighters on the ground became smaller and once we were 15 metres above ground, they grabbed their gear and trudged single file, disappearing into the thick black spruce forest towards the direction of the burned trees.

As we gained altitude, another clump of burned black spruce appeared amongst the muskeg. No smoke or fire was present, but we would still need to land and make sure the wildfire was out before we could move to the next start.

"XMA Two-Six, this is KNJ, you can check us off PWF 056 minus Bravo crew. We have the next smoke in sight. Couple one-tree wonders and no other resources needed. Requesting fire number," I transmitted.

"That's all copied. The next fire number is PWF 059," replied Two-Six. In a matter of minutes since dropping off Bravo crew, two other wildfires had been confirmed in the district. A fire flap, where multiple wildfires start at once, was happening, and we were in the middle of it.[7]

Getting the info ready for dispatch that mirrored the last fire, I passed on the white message and confirmed that we would be landing on fire 059. We found a suitable landing spot in a clearing less than half a kilometre from the fire.

"How long do you think it will take you to go fuel up and come back?" I asked Shane. He punched in the coordinates to Haig Base. "About a 45-minute turnaround so with fuelling up I'll be just under an hour," he replied.

"Right on. We should be faster than the other crew so come back for us then we'll go back for Bravo," I instructed. The fire was only a few hundred metres from where we could land and with the fire likely being extinguished from the rain, I felt confident that we could get to the fire, confirm its extinguishment, and get back to the helipad before the pilot could return. With an approximate hour turnaround time for the helicopter it was more than possible.

---

7   Each fire is numbered sequentially when it is located. So PWF 059 was the 59th fire in 2013 in the Peace River District.

After instructing the three firefighters in the back seat to grab the other chainsaw, two Wajax bags, and our hand tools, the helicopter landed, and we unloaded our gear. Kneeling just outside the pilot's door, I gave the thumbs up, and the pilot took his machine back into the sky. Having honed my sense of direction in the wilderness I was able to immediately start bushwhacking towards our goal.

The black spruce forest closed around us as we ventured from the openness of the grassy field. The grass turned to muskeg. Each step felt like walking on a giant spongy mattress where water seeped from the path. With every step we disturbed the patiently waiting mosquitos who in their thirst feverishly chased us. I was grateful that this species of mosquito was the larger, less aggressive one that I was able to ignore unlike their smaller, faster cousins that I have seen drive firefighters into panic. I was conscious to never eat bananas as it seemed to perpetuate the bloodlust of those insects. Gary and Julie immediately reached under their helmet liners and put on their green bug nets hoping to keep the bugs from their exposed skin.

Even with the landscape rather flat, it was easy to get turned around while walking through the thick vegetation. Being able to use the sun as a cardinal reference helped but my skill had come from years of experience – as well as being turned around myself in the past.

"I can smell it," Jesse said. It was a common tactic to use scent to locate wildfires with the wind as a reference point. The smell of freshly charred wet wood filled my nostrils momentarily confirming Jesse's claim.

Equipped with my GPS that I occasionally looked at to make sure the distance to our objective was getting shorter, we were able to make it to the wildfire in eight minutes from landing.

Jesse, Gary, and Julie all had the same smile on their faces when we arrived at the fire. My grin remained hidden under the thick moustache. Although there is nothing sexy about bushwhacking through wet muskeg after a storm to confirm a dead fire, it was still a blast to be revved up with adrenaline from watching the fires start and getting in and out of a running helicopter. Over my career, I had acquired the skill of cruising through the bush at a pace few could keep up with. Being 6"1' helped, but also the experience of planning my route and how to stabilize myself in the muskeg gave me the advantage. I always got a sick satisfaction of watching the city kids try to keep up while stumbling through the forest like fawns learning how to walk.

The clump of three, eight-metre-tall black spruce trees with 25-centimetre diameters had been burned leaving behind only branches and charred bark. The top few centimetres of forest floor the size of a large dinner table was singed black. The tallest of the scraggy black spruce had a crack that spiralled from top to bottom showing evidence of the lightning strike. With pines, the lightning shatters the bark, throwing shards of the trunk off the tree. These small black spruce that grew in nutrient-poor, cold conditions grew at a slow rate and even though they looked young, they had lived close to a century. Their growth rings were thereby tight, adding to the strength of the wood.

The only thing left to do was cold trail the fire. The four of us took turns chopping down the trees with a Pulaski. As each one fell, we ran our hands over the bark making sure no hidden embers remained. After 10 minutes of cold trailing, I was satisfied that the wildfire was out.

"XMA Two-Six this is PR201, check extinguishment of PWF 059 at 15:24 and we will be heading back to our pad," I transmitted.

We made our way to the helipad at a more leisurely pace and 40 minutes from when I declared the fire out, Shane was landing back at our helipad. We then flew back to pick up Bravo crew from the other fire.

Landing back at PWF 056, the rest of the crew loaded their gear and themselves in the helicopter. Once back in the air, dispatch asked us to return to Haig Base. Knowing that we had seen a third fire before the storm had passed, I requested to patrol to search the area that had been hit by lightning. After that request was granted, we found PWF 062 several minutes later.

This fire was like the previous ones – it had been hit by precipitation and was less than 0.1 hectares in size. After a kilometre walk down two seismic lines and through a thick black spruce and willow stand, we arrived at the fire to find three small smokes under a tall black spruce. Its branches were covered in thick needles and old man's beard lichen. Together these kept the rain from saturating the ground allowing the heat to smoulder in the dry carpet of muskeg. Everything not protected by the foliage was soaked, including us. By digging up the ground we were able to smother the smokes with ease.

Knowing there were other fires burning in the district, I kept the pace back to our helipad brisk which laboured everyone except for Brando. His long moose legs were perfectly adapted to stepping over anything that got in his way. From drop off to the return back to the helicopter was less than two hours. We flew towards Haig Base to hopefully grab some dinner before another dispatch.

On the flight, we passed by a helicopter lifting from a well site with its bucket attached. A crew of Helitack firefighters disappeared into a lodgepole pine stand below in search of their target. The subtle difference in the shades of green helped delineate forest types from our altitude. During a winter gig where I was cutting down and burning lodgepole pine that was infected by mountain pine beetle, I learned to spot different tree species from the air. The lighter shade of lime green that separates it from their spruce neighbours was enough for me to tell the difference. Once you can identify trees from the air you have a better idea of the ecosystems they grow in. For example, if you see black spruce and tamarack, they grow in poor-nutrient wet conditions, so you know that fire is going to be shitty.

"Man it's busy out there," I said into the headset to no one in particular.

"Yeah. On my way back to get you I heard another heli picking up some of your crew to get those fires we saw earlier in High Level," said Shane.

"Right on. I'm sure that won't be the last of them," I responded.

The rolling forest below mixed with creeks and dirt roads became familiar as we approached Haig Base. On our decent a small helicopter lifted from base and flew south. Over the radio, a wildfire ranger transmitted that he, James, and Popp were en route to a fire a few kilometres from base. That meant Alpha and Delta must have been the crews sent to High Level which made sense to me. Over the last two seasons, I had come to trust Jordan to fight fires without me and I had been leaving him in charge of the crew in my absence. With Farley's experience, bringing him with Jordan was an obvious choice.

At 18:00 we landed back at base, the perfect time to grab some food. The moment the blades stopped, we got out of the helicopter and raced to the kitchen. The eight of us inhaled our chicken and macaroni dinner knowing that it could be only minutes until we were called again. As I shovelled the last of the food down my throat, my cell rang, and we were asked to relocate to MATB to intercept another storm front that was steadily moving west and dropping lightning in its wake.

Five minutes after taking off we were dispatched to a smoke in the central area of the district by the Notikewin Lookout, the same lookout where Cavelle and I spent time last summer. We were

45 minutes from the smoke, giving my mind time to reminisce. I had been so busy I was unable to spend much time with her and momentarily wondered why I was living where the civilized world ends. My life had become the job. A memory of Christmas Eve engulfed my thoughts. While in Australia I took a fire call on December 24 and was hoping to overnight so I could say that I fought a wildfire on Christmas. In my hubris I forgot how lonely that must have been for Cavelle. I knew if I kept this job, I would lose her. But this wasn't the time to be distracted so I snapped my attention back to the wildfire I was about to fight.

There was no smoke visible when we arrived at the coordinates. However, the blood-red retardant the tankers smashed the fire with gave away its location. We landed in a well site close to the detection and were able to find the lightning tree with ease. These where some of the largest lodgepole pines in the area that lined the Notikewin River. It was nice to get a reprieve from the muskeg and swamp fires and walk through a forest that most could only picture in their minds. Lush honeysuckles, snowberries, with the entirety of the forest floor covered in feather moss. The fire was out, and the only action required was to fell the massive lodgepole pine that was blackened by the strike.

We flew back to the MATB and refuelled. 16 minutes later we were dispatched to two smokes that had been called in by the assessment helicopter. It was 21:12 and legal down was now a priority.[8] With a 30-minute flight back to Haig Base and legal down at 23:15, there was realistically only time to fight one of these fires. Upon arrival at PWF 069, there was no evidence of a fire and without a close landing pad and with daylight running out,

---

8   Visual flight rules (VFR) state that helicopters must be grounded before legal down, 30 minutes after sunset.

I made the decision to see if we could instead find PW F070 that was located 11 kilometres away.

I had never been on five fires in one day and we were so close I could taste it. Arriving at the coordinates for PWF 070, a large lodgepole pine stood out amongst the others with grey smoke rising from its base. The tree was on the edge of a dirt road surrounded by well sites making landing and access ideal for the time frame we had left.

"Okay, let's make this fast," I said to the crew. We dropped in altitude towards the bare-earth well site and like the four times earlier, unloaded from the helicopter and made our way towards the smoke.

"Man it's good to be back working with you bud," Brando said as we led the crew down the dirt road.

"Yeah buddy. It's great having you back," I said. In a couple minutes we cruised the 200 metres from the helipad to the coordinates and the fire was easy to spot from the roadside. The base of the tree was split open with shards of wood scattered around the base. A 1.5-metre-long chuck of wood was blasted out of the trunk from the lightning strike that had rippled through the tree. The ground around it smouldered but was contained to a metre or so from the tree base. White chalk powder was attached to the bottom of the pine and on the smouldering ground.

"What the hell is that?" Couse asked.

"Looks like someone used a fire extinguisher on it," Brando said. "Probably someone driving by saw it and tried to put it out."

"Let's get moving. Line up and use our lids with the ditch water," I said. Each of us pulled out our helmet liners and unattached the peltors and face shield. One by one we passed our helmets back and forth along our human chain. With each helmet full of water the ground became saturated, and the smoke disappeared. "Cold trail time and let's get outta here," I instructed. We dug our bare hands into the earth feeling for any sign that the fire was still alive.

It was 22:30 with little time left if we wanted to make it back to base by legal down. "What do you think Brando?" I asked my friend.

"Good enough for the woman I sleep with."

"Nice," I said with a laugh. "Let's get back to the pad."

I grabbed the long chunk of wood that was blown apart from the tree and leaned it over my shoulder. It would be an awesome reminder of the day and would be a great centre piece on my wall at home. When we approached the helicopter, the sun was no longer visible from the road. Pink and red hues from the setting sun mixed with the darkness of the cloudy sky above.

We flew back to Haig Base in silence, each firefighter lost in their own thoughts about the day. As we descended to ground one final time it was 22:58, beating the darkness by 15 minutes. The sun was hidden behind the horizon with only remnants of its lingering light.

I met with my other leaders and found they had successfully fought and extinguished their fires. Even though we were not working directly together as a Unit, we had become skilled individually to be utilized on several fires at once. In less than eight hours we had crushed seven wildfires.

The following day we loaded into three helicopters and flew in tandem to three more fires in the High Level District. It was a rare experience to have my whole crew flying beside me, each split up in a helicopter to my nine and three o'clocks. In a culture where fire numbers are bragging rights, 10 fires in two days was a respectable number to add to the stat sheet.

In total, we had fought 15 fires so far this season, over twice as many as the previous one, and the summer was just beginning. But unlike last year, the fires would keep on starting.

Dispatch Notes

1342 dispatch → 57 15.965 116 13.002
1347↑Haig Base→ PWF 056, BD 55 north of detection working on 4 fires
1445 7 Smokes visible, unable to action due to storm
1459↓ PWF 056, dropped off 205, 206, 207, 210 (Marco, Brando, Des, and Couse)
1504↑PWF 056→PWF 059
1512↓PWF 059 with 201, 211, 212, 214 (Larson, Jesse, Gillis, and Thirnbeck)
1520 Arrive at PWF 059, 0.17 mile walk
1524 EX PWF 059
1606↑PWF 059→PWF 056
1611↓PWF 056
1613↑PWF 056 with all 8 members →Patrol Local
1624 Located PWF 062
1632 ↓PWF 062
1659 Arrived at PWF 062, 0.51 mile walk, IC 201
1717 EX PWF 062
1740↑PWF 062→Haig Base
1801↓Haig Base
1825↑Haig Base to MATB
1830 Dispatched to PWF 067, 45 miles away
1921 ↓PWF 067, 0.17 mile walk
1931 Arrive at PWF 067
1945 EX PWF 067
1954↑PWF 067→MATB
2019↓MATB
2035↑MATB→PWF 069/PWF 070
2112 Arrive at PWF 069, cannot locate
2127→PWF 070
2133↓PWF 070, 200 metre walk
2141 Arrive at PWF 070
2228 EX PWF 070
2231↑PWF 070→Haig Base
2258↓Haig Base

* ↑ lifting from ground, ↓ landing, → going to, EX fire is extinguished. In typical Canadian fashion the notes are a mix of metric and imperial.

15

GP 43

"If Harold Larson says he's got this fire, then he's got it."
- Ben Jamieson, Provincial Unit Crew Coordinator

I sat in the passenger seat of the Bell 212 helicopter as we flew over the asbestos forests in the Grande Prairie District. The giant woodlands of trembling aspen and balsam poplar had gained this unfortunate name in wildfire circles due to the deciduous trees that dominated much of this landscape which never seemed to burn. We were sent on a patrol to scout out the western area of the district south of Peace River that bordered British Columbia. On large fires I have used stands of deciduous trees as natural fuel breaks from their highly volatile coniferous neighbours. Seeing the never-ending leaved forests for the first time, I understood why the area south of my home district burned far less each season.

The seven firefighters of Bravo and Charlie crews sat in the back seats of the machine. Brando, DK, and Marco each had headsets, allowing them to hear dispatch and take part in the conversation. Jesse, Julie, Dezzie, and Couse sat silently with their peltors over their ears, unable to hear anything but the muffled rumble of the rotor blades.

It was July 15. During the two weeks since the fire flap, we had been exported from Manning to Fort Vermillion, one of the most northern bases in the High Level District. After spending four days at that base, we chased the fire hazard south to the Graham Base in the Grande Prairie District.

The day after our arrival, we fought a small fire in an old cut-block. The fire wasn't hard to control. However, the high drought codes and the endless slash and wood chips that covered the forest floor made the extinguishment of that fire time consuming.[9] In those conditions, fire burns deep and hides in areas that require extensive searching and digging to find. After two days of hard mop-up, the fire was in search-and-destroy mode, so I left Jordan in charge of 11 firefighters while I took the remaining seven to man-up our helicopter for initial attack duties.

As I scanned the horizon for any hints of fire, my train of thought was interrupted by the voice of a wildfire ranger in the assessment helicopter. Over my headset he said, "XMC565 this is Charlie Whiskey Echo. We have a small smoke in sight."

I scribbled down the coordinates into my notepad then transferred them into my GPS. It showed us 10 minutes away from the fire and close to where the rest of our Unit was working. Being the nearest resource to the fire, our helicopter was rerouted in that direction.

---

9   Drought codes are used to help predict how deep a fire will burn into the ground by combining the precipitation, and temperature, into an equation.

"Peace Delta Leader this is Uniform Whiskey X-Ray," I transmitted, hoping I was close enough for it to reach Jordan on his radio.

"Go for Delta Leader."

"How's your fire looking?" I asked.

"Pretty dead. Only found a handful of spots so far."

"Perfect. There's another start west of your location along the road. We are nine minutes back. Get everyone and drive west. We'll meet you there."

"That's all copied."

The asbestos forests transitioned into a mix of white spruce and lodgepole pine. Gravel roads intersected with cutblocks revealing themselves amongst the trees with every passing moment. Light-grey smoke appeared from a cutblock several kilometres in the distance. Three minutes from the coordinates we flew over the fire that the rest of our Unit was working on. The blue and red helmets of the firefighters below stood out amongst the charred remnants of the old fire. They weaved through the blackened stumps towards the three white trucks parked alongside the gravel road that bordered the fire.

At 15:10 we arrived at the new start. Our pilot went into assessment mode and flew the helicopter in counter-clockwise circles so I could see the fire from all sides. It was a smouldering ground fire half the size of a football field, and it was creeping through the slash surrounded by trembling aspen. These 25-metre-tall trees had snow-white bark and small heart-shaped, emerald coloured leaves that fluttered in the breeze. They had been left unharvested because of their undesirable wood quality.

After the first circle around the fire, I could delineate that three-quarters of the fire was smouldering while the rest appeared to be an old burn previously circled by a dozer guard. The south flank of the fire was contained by the gravel road that the rest of the Unit was coming from nearly a kilometre to the east. A full dug-out[10] 180 metres to the north of the fire's edge would be a perfect water source to utilize. With our Mark IIIs and by running hose directly to the fire perimeter we could establish a wet-line around the fire without any foreseen complications.

Calling dispatch, I cancelled the incoming tankers and requested a dozer group to create a fuel break around the slow moving fire. A Helitack crew and a Rappel crew arrived by helicopters and joined the show. With the combined strength of 32 firefighters, three helicopters, and three dozers, we made quick work stopping the fire.

By 17:00 a dozer guard was built around the smouldering stumps and slash in the old cutblock. This fire had reignited from a previous small fire that was thought to have been extinguished a week before. Cutblock fires burn deep, and an ember must have escaped the mop-up process and reignited in the heat of the day.

A backhoe that was rehabbing the week-old dozer line was used after the dozers had created the new guard. The giant yellow machine crept along the bare earth using its powerful bucket to push over danger trees. With help from the mechanical dinosaur, our fallers could concentrate on extinguishment duties.

The fire was now contained to smouldering slash that covered the old cutblock so I felt comfortable releasing the two initial attack crews that had come to help. After calling dispatch and changing the status of the fire to *Being Held*, Aussie Rob, one of the Rappel crew members, informed me that their helicopter was US (unserviceable) and needed their engineer to fix it before they

---

10 Dug-out or borrow-pit. To build roads in certain areas, earth would be dug at intervals used to raise the road above the groundwater. These dug-outs are usually rectangular in shape, varied in size and depth, and all filled with water creating miniature ecosystems.

could fly anywhere.[11] The Helitack helicopter would be flying back to base to pick up an engineer to take him to the fire where he could work on the downed helicopter.

"Well, looks like you guys are helping out until you can get your bird back into the sky," I joked with Rob.

"No worries mate. Glad to help out."

Satisfied that everyone was where they were supposed to be, I joined Jordan mopping up the fire. Finding a nozzle to use was difficult after the crew had established themselves on the line, so I helped dig up spots with the shovel-of-a-thousand-truths. Most firefighters, especially the newer ones, have a proclivity toward the Pulaski as the hand tool of choice. However, I was a natural with a shovel and I loved ripping apart anything that was burning with this hand tool that had become an extension of my body.

Feeling my cell vibrate, I walked away from the noise of blasting water to answer the call from dispatch. Time to go to another fire. Since the Rappel and Helitack helicopters were not available we would have to step in again for initial attack. Too bad for them, but I didn't feel one ounce of pity. Going to fires becomes competitive and the excitement of new starts trumps other crews' hurt feelings. Another fire had started 50 kilometres south of us and we were once again the closest resource.

"I have to get going to another dispatch. You good to take this fire over?" I asked Jordan.

---

11 Wildfires are classified as either: OC – Out of control; BH – Being Held and will not grow more than 10 percent or beyond predetermined boundaries; UC – Under Control and will not grow; EX extinguished; or TO – turned over to landowner. These classifications help to determine strategies, tactics, and resources required. The provincial goal is to have all wildfires BH by 10:00 am the following day from its discovery.

"Yeah man. We got this under control. Go get another one," he replied.

I tilted my head and spoke into my radio mic. "Time is 18:10 and Peace Delta Leader will be taking over command of GWF 042. Peace Charlie and Bravo crew make your way to the machine. We have another dispatch."

The firefighters that were called handed their hoses to the other members, gathered their equipment, and hustled to our helicopter parked opposite the road to the fire. We loaded up our gear and at 18:20, lifted into the sky, watching the crews on the ground working around the trembling aspen until they disappeared from view.

Once we gained elevation, the westward view towards British Columbia was concerning. Far off on the horizon, dozens of smoke columns merged into a black haze. A wildfire, thousands of hectares in size, was ripping towards the Alberta boarder. The dark cauliflower pillars rose hundreds of metres with mushrooming tops that were sheared off from the change in atmosphere. The thick smoke covered the sun making the day appear like dusk. The fast westerly winds pushed the monster called the Tumbler Ridge Fire through anything that stood in its path. Tightly packed mature spruce and pine forests, pocketed with well sites, stood little chance

---

* For you fire nerds here is the weather readings for that day from a nearby weather station.

Time	Max Temp.	Low RH	Wind	FFMC	DMC	DC	BUI	ISI	FWI
14:00 - 16:00	31.3C	15%	10 kph West	94.8	82	317	100	16	42.6
				Extreme	Extreme	Very High	Extreme	Extreme	Extreme

as the Tumbler consumed fuel faster than any entity could contain it. We flew in silence for several minutes watching the spectacle that few in the world except the nine of us were witnessing.

The silence was broken by the radio chatter of the Bird Dog directing a group of tankers over our next objective, GWF 043. The forest gave way to a deep river valley. A turquoise river flowed between the steep slopes that lined the crystal waters. We were only several minutes back from the fire but couldn't see any smoke in the direction of our dispatch because of the contours of the river valley. We transmitted to the Bird Dog that we were five minutes back. He asked us to hold our position until the air tankers could finish their retardant drops. He relayed that the fire was about 10 hectares in size and that they were trying to contain it on the hillside to keep it from reaching the forest above. There were also running retardant strips along the crest of the hillside hoping to contain it in the river valley.

"Copy that, standing by for clearance into 043," I responded on the radio.

For 20 minutes we flew in wide circles staying back from the air tanker action. We watched the Tumbler Ridge Fire continue to consume the forest in the distance. Over the radio we heard that the fire had crossed the invisible Alberta border and was labelled GWF 044.

"Well looks like the BC Wildfire Service has the east flank taken care of," I said to the pilot. Working in a district that bordered BC, this was a common joke when referring to a wildfire that would burn east into Alberta forcing us to take over its eastern flank.

Once we were cleared into the airspace by the Bird Dog, we lifted over the ridgeline and flew low-level towards the fire. It was burning mid slope to the top of a ridgeline that faced south towards the crystal blue Narraway River 120 metres below. Massive trees torched as the fire burned along the hillside. The forest was an uncommon mature black spruce and lodgepole pine mix that was thick from crown to crown for several kilometres in all directions. We flew low-level over the ridgeline looking for any opening to land in. The thick canopy layer was closed in all directions for kilometres. We found a grassy field several hundred metres above the ridgeline, but the thought of landing there had to be abandoned. Without solid anchor points, I couldn't risk hiking blindly through that forest with an uncontrolled fire below.

Unable to find a safe helipad above the ridgeline, we landed on a sandbar on the river below. The steepness of the slope mixed with the loose rocks making it too unsafe for me to realistically get my crew up by foot. There was nothing any ground resource could have done that evening to stop this fire. The fire was a constant HFI 3 with the massive trees torching every couple of minutes.[12]

The Rappel helicopter appeared over the ridgeline into the airspace above us. "IC 43, this is Rap 10. Requesting permission to rappel in and build a pad," came the voice of the spotter over the radio. It was 20:15, meaning we only had an hour until we would have to fly back to base to beat legal down. Building a helipad in that thick mature timber within an hour would be impossible. With the combination of air traffic and the fire conditions, the safety measures were not in place for me to make that call. The tankers and helicopters also needed the airspace to continue to work on the ridgeline.

"Request denied. That will be your objective first thing in the morning. Look for a spot off the southern ridgeline and prep for tomorrow," I responded.

---

12 HFI – Head Fire Intensity is the measurement of how a fire is burning: 1 being the lowest, smouldering, to 6 being the highest, an out-of-control wildfire where all tactics will fail. HFI helps determine strategies and tactics.

We climbed up the opposite slope and watched the aerial display. Convairs and Electras, massive, fixed-wing tankers, hammered the ridgeline with their blood-red payloads hoping to stop the fire from spreading into the forest up on the flats. Between drops, the two helicopters lowered their orange pumpkin-shaped buckets that hung 30 metres below them from cables, into the river, filling them with water. They hit the smaller targets by releasing the water with remarkable accuracy, trying to hold the fire on the hillside.

Before it was time to leave, I received a call from the duty officer who was also the provincial Unit crew coordinator. So, kind of like my boss's boss in a sort of way. In my two years running the Unit, it was my first time speaking to him.

"I looked you up in FIRES.[13] Seems like you have seen your share. Think you can take this fire on?" he said. The anticipation created from the Tumbler Ridge Fire had tied up the local resources including the Grande Prairie overhead staff. I was the best option to stop this fire.

"Yeah, got this no problem. Only issue is no safe landing area close to the fire. Going to need Rappel for the morning to build a pad," I answered.

"I can't believe you just said that," he said with a laugh. He confirmed that I would be IC of fire GWF 043 and said to start figuring out what resources I needed to stop this fire.

The weather forecast for the next day showed similar conditions to what we had experienced. At 20:30, the temperature was 31°C with a relative humidity (RH) of 32, close to crossover. We

---

13 FIRES is the program that records each firefighter's wildfires, position, and hours spent on each one. By that time, I had 189 fires and 3000+ hours attached to my name, not including my five years in BC and one year in Australia.

would have to get an early start the next day to beat those conditions. With that in mind I was confident that we would have a window until the early afternoon to get this fire under control.

I had a Rappel crew at my disposal and for the first time I could use them for what I was told was their strength: rappelling in and cutting a helipad. With daylight running out, that would be the plan first thing in the morning. With a helipad constructed we could fight the fire up close and personal.

16

JUST ONE MORE HOUR

"This is bullshit."
– Martha

The eight of us looked up at the two helicopters that were bucketing on the fire that was slowly spreading along the western slopes boarding the river valley 100 metres above. The "whoop whoop whoop" of the large metallic blades mixed with the roar of chainsaws that echoed throughout the river valley. The ambient flow of the river did little to ease our nerves. The four-hour estimated time for helipad construction by the Rappel crew on the south end of the fire had come and gone. It was noon and all we could do was sit and wait on the rocky riverbed that lined the turquoise river that flowed around us.

The seven other firefighters were hand-picked to be with me from all the sub-crews. Jordan, Brando, and Farley because of their experience and wildfire knowledge, the Fulthorpe brothers because of their physical strength, Soso, because I trusted working with him, and Nate. The year before Nate started on the Unit, he had lost a substantial amount of weight after moving to Central America and focusing on ju-jitsu. He was the perfect hose donkey.

During the previous evening, tankers had dropped retardant line after retardant line along the ridgeline, containing the fire to the slopes below the forest that had established itself on the mountain hundreds of years ago. The retardant was holding but without support from the ground, it would only be a matter of time before fire from somewhere would find its way through to the forest on the other side.

Throughout the morning hours the fire had remained tame. Light smoke rose through the canopy exposing the fire's growth, but the flames remained hidden in the understory. The rapid torching of trees we witnessed the prior evening was non-existent. There was still time for us to fight this one. The smoke was concentrated on the upper third of the hillside that comprised a mix of mature lodgepole pine and white spruce. From the ridgeline to the river, the thick canopy became sparse to non-existent as the slope transitioned to dark-grey shale as it approached the river bottom where we patiently waited. The ridgeline above was oriented north/south and was approximately one kilometre long before each end turned west following the water that had carved out the river valley.

Where the south end of the ridgeline turned, a giant rocky outcrop jutted out into the open sky. The 10-by-30 metre curved and jagged outcrop that had been formed by thousands of years of erosion, dropped straight down to the river below with a steep, almost completely shale-covered backside. The natural barrier would be the best place to hold the fire on that flank if needed.

At the opposite end of the ridgeline stood a smaller, yet equally impressive rock face. Behind it, a smaller river drained into the main river creating another natural break to potentially contain the fire.

The fire burned in the middle third of the ridgeline slowly creeping towards each rock face. The south edge of the fire was only 300 metres from the partially completed helipad.

"Rap 10 this is IC 43," I radioed to the Rappel spotter in one of the two helicopters that was bucketing on the fire.

"Go for Rap 10."

"Yeah, just wondering about an ETA on the helipad?"

"It's coming along. I would say about one more hour," the spotter transmitted back.

I closed my eyes and let out a long sigh trying to release some of my frustration. "Copy that. Standing by," I transmitted. If I couldn't get my crew close to the fire before peak burning time, we would miss our opportunity to safely action the fire.

"Fucking Rap," said Brando who was standing between Jordan and me. "That's the third time we have heard that nonsense. How long does it take to make a helipad?" Brando shared little love for the Rappel program. He had spent his career in Peace River, a district that seemed to shy away from importing Rappel crews into the area.

Being a product of the Peace River system, I too held little affection for the Rappel program. My feelings were reinforced by the limited time I had seen them fighting fires. Of the times our paths had crossed, they barley give my crew the time of day. Knowing the foolishness of my prejudice and wanting to avoid childish revelries, I was willing to give this crew the benefit of the doubt. My confidence, however, waned with every passing hour my crew remained idle waiting for the helipad.

Letting our impatience get the best of us, we began to scour the hillside for a way to the top of the ridgeline. To the south where the vegetation was thickest, the slope was too steep for anyone to climb safely, especially with gear and no anchor point. Adam and I made it 60 metres up the shale slopes before realizing the futility of our efforts. For every metre up, we would slide down two-thirds that. From the river bottom, it was deceptive how steep the hillside was. I always kept the well-being of the crew at the forefront of my decision-making and the idea of mountaineering was thereby abandoned.

As tempting as it was to attempt a pump relay system from the endless water source of the river, no area could be found that could hold a full bladder mid slope. I also didn't want my crew scaling up and down the dangerous rocks.

We discussed a small grassy field on the north side of the mountain for a landing spot, but with the grassy field surrounded by continuous mature timber without any escape routes, that idea also had to be discarded. Our only option was to wait for the Rappel crew to do their job and finish the helipad.

"IC 43, this is Rap 10, check me flying back to Graham base," came the voice of the spotter over the radios.

"Why are you going back to Graham? Kakwa tower is half the distance," I transmitted knowing that the helicopter was due for fuel and that the Kakwa fuel cache was much closer than that base.

"Going back to camp to grab some lunch for the crew," the spotter transmitted.

"Pardon me?" I radioed back.

"It's taking longer than we thought. Check me heading back to Graham." With that, the helicopter veered northwest and began its flight back to base. It would be over one and a quarter hours before that helicopter would be back, taking away from precious time it could have been bucketing.

"That's all copied," I transmitted after a long silence. Fighting the urge to throw my radio into the river I instead placed it back into the leather holster on my belt.

"What type of crew goes to a fire and doesn't bring food?" Jordan asked rhetorically.

"Apparently they do," Brando answered.

We paced anxiously along the riverbed waiting for our turn to get into the action. It had been almost 30 minutes since the Rappel spotter had left and since his departure the familiar sounds of chainsaws had ceased from the helipad location. Being on the end of a chainsaw for hours on end is tiring work. I respected that breaks are needed but with the heat of the day increasing, so too did the fire's rate of spread and intensity. The occasional tree was beginning to torch, and I was beginning to worry about the crew above. Calculating the indies[14] and using yesterday's fire behaviour as a benchmark, I knew we were running out of time.

"Anyone from Rap 10 got a copy?" I radioed.

"Yeah this is Munday. How's it going mate?" Munday transmitted.

"Good to hear you Rob. How's it going up there?"

"We're getting there. Just taking a breather," Munday transmitted back.

I did not want to push a fatigued crew but with the fire activity picking up there was a job that needed to be done. With two chainsaws and six firefighters they should be able to push it for a couple hours more and be fine. In truth I really didn't give a shit about their fatigue management at that moment.

"Hey buddy, I know you guys are probably tired but the fire is getting close and I would suggest you get back to building that helipad so we can get you out of there. Hope you guys have an escape route planned up there," I radioed, hoping that my mention of an escape route would motivate them.

"Yeah copy that," Munday transmitted.

---

14 Calculating fire behaviour is a science that requires combining temperature, relative humidity, wind, precipitation, atmospheric stability, topography, aspect, slope, and fuel type as well as other factors that make up books themselves.

My crew paced along the riverbed, mumbling to each other about the situation. Ten minutes passed and the roaring of chainsaws still did not echo throughout the river valley. The helicopter that was assigned to my crew was bucketing on the south flank of the fire closest to the Rappel crew. The one bucket was not enough and the fire on the south flank continued to spread. It was 14:30, and the time when we could get established on the fire was drawing closer to the point of no-go.

"Hey Rob. I would highly recommend your break time to end. The indices are picking up and the fire activity is increasing by the minute. Your helipad is going to be compromised within the hour," I radioed. The frustration that was surging through my body was now being conveyed through my tone over the radio. Their only job was to build the helipad. I wasn't up there with them so I had to give them the trust that they were doing the best that they could. I was powerless on the riverbank and needed them to hurry the fuck up.

"Yeah mate. All good," came Munday's voice over the radios. Still no sound of purring engines or cutting wood. Several minutes passed.

"Yeah Harold this is Munday. We ran out of fuel for our saws," transmitted Munday.

"Oh my God," I said to no one in particular. I was now more frustrated with the Rappel spotter who had had not asked for help and left his crew. Precious chainsaw and bucketing time were never going to be given back because of that decision.

"No worries Rob. We have lots down here. We will get it slung to you right away," I radioed. I instructed my pilot to lower his bucket to the riverbed where Jordan placed our two combie cans of fuel in the bright orange bucket. Within minutes, the fuel was slung up to the Rappel crew and the sounds of progress echoed once again throughout the river valley.

At 15:30 trees began to torch at frequent intervals along the hillside. The retardant lines that had held throughout the night and day were failing and the fire breached the north flank where it had been tied in. Without being able to action the fire on foot and without a helipad for us to use, I knew that our window to fight this fire was gone. The Rappel spotter and the other helicopter had returned and even with both helicopters bucketing on the escaped fire, containment was not possible. Hearing from our pilot that he was unable to stop the spread on the ridgeline, I requested air tankers for support.

An hour later, the helipad was still under construction and the fire had burned far enough south to threaten the unfinished helipad.

"Rap 10. The fire is getting pretty close. It's now or never time," I radioed.

"Copy that."

Ten minutes later they finished the pad and got flown out of the fire to sit with us at the river bottom. Soon after they landed, an increase in smoke developed below the pad making it unusable. The pad took eight and a half hours to complete. That's right, eight and a half hours. The fire activity had increased to the point where it would not have been safe to put crews on it. It was now in fact a no-go. Without being able to get to the fire's edge, the only option I had was watch the incoming airshow.

When the Rappel crew got out of their helicopter all of them except Rob stripped off their clothes and jumped into the river. The six naked men splashed water on each other while some of them wrestled, trying to submerge one another.

"What the fuck is going on?" Brando asked me.

"Guess they're just a close crew," I replied.

"Way too close," said Brando.

The fire held well on the south flank. The constant bucket drops from the two helicopters kept the fire intensity low enough to slow its growth. Without added support, the fire on the opposite flank ripped along the hillside. Lucky, at 16:40, two Convairs and a group of 802s arrived and continued to tie in the ridge with retardant. They also did an excellent job of tying in the ridgeline with a massive rock face on the southern slopes

Calling the duty officer, I updated him on our progress or lack thereof and what we would need the following day. He wasn't pleased with us not being able to action the fire, but I had to reinforce that it couldn't be done safely. With the helipad now done we could hit the fucker hard the next morning before the fire activity picked up.

Later that evening, Hotel Zulu Foxtrot, an MD900 helicopter, came to join in the bucketing show. This machine was flown by a pilot that I worked with a lot when I led a Helitack crew. This machine was unique, not because it didn't have a tail rotor, but because it was formally owned by Tiger Woods. He remained bucketing until we left at the end of the day and was given to us as a resource to use to fight this fire the following day.

Upon departure at 20:50, we did a final flight over the fire and saw that it had spread from 5 to 10 hectares. The fire remained on the hillside and was burning 100 metres from the south helipad. Four hundred metres of unburned forest remained between the helipad and the southern rock face. The retardant was in place and would hopefully hold throughout the night.

From the helipad to the northern rock face, the fire had burned patchy along the hillside. Where the slopes narrowed into chutes, the fire burned hot but stopped at the retardant lines. It had gained momentum as it cooked off the northernmost part of the hillside and spotted across the smaller river valley 200 metres to the north. That spot fire had raced up the opposite slope on the northern gully and stopped as soon as it hit the ridgetop. It was inaccessible by foot and would require a crew to rappel in to create another pad on that side of the gully. I did not want anyone walking blind into the fire through kilometres of continuous forest, so Rappel was once again our only safe option. Rap 10 was due to be timed out and a new Rappel crew would be completing that objective. Hopefully their helipad building skills were superior to that of their peers.

The excessive amounts of smoke created from the Tumbler Ridge Fire drifted in the air. We flew home low over the treetops watching the crimson skyline from the sunset.

Even though we could not get on the ground we felt that this fire was still in our grasp. With the south helipad built and the ridgeline 90 percent tied in with retardant, we would attack by ground the following day.

17

IS THIS THE END MY FRIEND?

"Sorry Joel, this isn't amateur hour."
– Harold Larson

Today we'd hit the fire hard. Jordan, Brando, and I developed a plan to move firefighters to the top of the ridge – and put in place the pieces we would need to fight this fire. Over the last day and a half of not directly attacking fire 43, my frustrations grew. But I could not let that blind me from keeping everyone safe.

Our strategy was to hold the fire below the ridgeline by tying the north-to-south flanks with pump and hose. The possibility of finding water above the ridgeline was low. We would use the helicopters and their buckets to fill up portable bladders from the river's endless water source. The fire on the hillside could consume whatever it wanted if we could hold the ridgeline.

With multiple tasks running simultaneously, it was imperative that everyone understood their assignments and where everyone else would be. Calculating how the fire had previously burned and with the forecasted weather, I knew we had until around 16:00

before the fire behaviour would become unmanageable. As much as this fire needed all the firefighters we could throw at it, I did not want more firefighters on the fire than the helicopters could extract. Having Echo crew as lookouts and Charlie crew running gear helped utilize the firefighters that I could not deploy on the fire.

After I was satisfied that everyone knew their objectives for the day, I walked to the helicopter with the seven others from Alpha and Bravo crews. Our objective was to land at the south helipad that was built the previous day and start containing the fire from that flank. Once again, I choose Bravo crew so I could utilize Brando's experience.

Our pilot had timed out the night before, so we were sent a replacement. As with all pilots, I immediately put my trust in him and listened to his long and detailed briefing.

As the crew climbed into their seats, I jumped into the front seat and strapped myself in. Brando, Farley, and Marco sat next to the doors and put on the three available headsets. Soso, Popp, Desiree, and Bonesaw filled in the middle.[15] As the rotors began to spin, the pilot punched buttons on the GPS mounted on the dash between us. Several minutes later, he continued to button mash, but the screen remained off.

"Need some help there?" I asked over the intercom after another minute of watching him fumble around with the GPS.

"Ah yeah. I can't seem to remember how to turn this on," the pilot replied. I reached over and held down the power button on the side of the GPS. The screen turned from black to blue as

---

15 Nigel, the younger of the Thompson brothers, asked to be referred to as Bonesaw. Like the Warhammer, he remains a conundrum to me. After spending years on the Unit, he gave up working and continues to travel the Earth looking for the meaning of life.

numbers and coordinates lined up on the display. "Okay, so how do I plug in the coordinates?"

You must be fucking kidding me. "Do you not know how to use this GPS?" I asked.

"Not this one. Been a while since I've done this."

Letting out a deep breath, I scrolled to the GWF 043 coordinates and leaned back in my seat. I could tell that this pilot was going to be garbage but there was a job to do, and you do not always have the luxury of a skillful pilot.

"Anything else you need help with before we take off?" I asked.

"Nope. Should be good to go now."

"Should be?"

The pilot lifted on the collective until the helicopter raised off the ground and started towards the fire. The horizon was thick with black smoke from the Tumbler Ridge Fire that was still out of control and had grown exponentially over the past three days. For the first couple minutes of flight, I taught our pilot how to use the radios and the etiquette needed when talking on the radio. After 15 minutes, the smoke in the air created a light haze, almost like fog, that made the pilot lower the helicopter's altitude as not to lose sight of the treetops. Being less than a couple hundred metres from the canopy made the breakneck speed at which we travelled that much more apparent. My eyes strained as I looked for any changes in the topography that could take us by surprise.

"HG. Never trust them," the pilot said breaking the silence. I continued to look ahead. "Had an HG engine explode while I was flying. After that I won't even have an HG microwave."

I remained silent, concentrating on our flight path, now with even less confidence knowing that our pilot had crashed a helicopter, even if, according to him, it was due to no fault of his own.

Thirty minutes after liftoff we approached GWF 043. The smoke cleared over the rolling hills of continuous trembling aspen and balsam poplar. These trees soon gave way to the large coniferous forests that lined the steep rocky canyons created by the turquoise river that flowed through it. The fire had not spread during the night. Asleep from the coldness created by the absence of the sun it instead just endured. Hundreds of smoke tendrils reached through the canopy searching for the heat it needed to wake from its slumber.

We circled the south helipad checking for clues on how best to work that flank. One hundred metres downslope from the helipad, sunlight reflected off the forest floor showing there might be water to use on the ridgeline. "Hey Brando, looks like we might have a water source below our pad. Go check it out for us, eh? Who do you want to take with you?" I asked over the intercom.

"Yeah man, I see that too. I'll take Dezzie," Brando responded.

Continuing to circle the helipad, I examined the work that had been done the previous day. Dozens of trees had been felled and bucked into smaller pieces that lined the perimeter of the pad. It was going to be a tight fit. A lot of work had been done but with the haste needed during construction, the size was less than ideal, which made me nervous about landing there. We were trained to make helipads at least twice the length of the rotor blades so the pilot has room for error, especially when the machine must drop vertically through a forest canopy. Gaining lift without moving forward is also a skill pilots acquire through experience. This helipad barely made the minimum specs but ultimately, the decision to land would be up to the pilot.

"You feel comfortable getting us in there?" I asked.

"Yeah…"

"Okay, let's do this."

After the shaky assurance and seeing that the fire was at low intensity, I called into dispatch and relayed that we were going to land. The pilot slowed the helicopter and began the descent towards the opening in the canopy.

The helicopter lowered into a hover directly over the centre of the helipad. As the skids lined up with the tops of the large conifers that outlined the helipad, the eight of us collectively held our breath and remained motionless. Ever so slowly, the helicopter began to sink towards the earth 30 metres below. Our view of the cloudless horizon was replaced by the spired tops of the evergreens. As the main rotor blades passed under the line of treetops, the machine suddenly dropped 10 metres and shifted to the left, the spinning blades brushing only metres from the giant trees that now encircled us.

Loud sirens blared in our headsets and throughout the helicopter. Warning lights flashed across the dashboard.

"Oh shit! Oh shit! Oh shit!" the pilot repeated as he struggled to control the helicopter that was plummeting to the ground. Dropping below the canopy, it lost its lift created by the wind and was now, nowhere close to being in control.

Brando, having been in thousands of landings before, knew something was wrong. Several years earlier, he had been in a helicopter that had blown an engine and had to make an emergency landing. But this was the first time he felt the unfamiliar emotions associated with an imminent crash.

Oh please not here. Dammit, I'm right in front of the transmission.

Over the years he had heard horror stories about helicopter crashes and how the engine can come through the back seat and destroy who or whatever was sitting there.

Popp looked out of the window and smiled at how fast the trunks of the trees were moving vertically beside the machine.

Farley clutched the seat getting ready to absorb the ugly landing.

Desireé, hearing the sirens through her peltors, turned her eyes to Brando for comfort but instead saw an unfamiliar look on his face.

My only thought was that I had failed my crew. We were going to crash in an area that couldn't fit another helicopter and evacuating anyone that might survive would be complicated. I also knew that when that style of helicopter crashes, the people in the front usually do not survive. The warning signs were there that this pilot was trash and I had ignored my intuition.

So this is how it ends. I closed my eyes, took in a calming breath, and braced for impact.

18

DELTA CREW

"It's better to ask for forgiveness than permission."
– Jordan Sykes

The helicopter holding Delta crew circled above the northern point of the ridgetop looking for the best area to make a helipad and establish a bladder site.

"Looks like that scorched off nicely," Jordan said over the headset to his crew. The fire had burned hot up the northernmost slopes and cooked off a rocky area that looked big enough to complete their objective. It was still too small to land their helicopter but a grassy meadow to the west had been scouted the day before and would be the spot where they could exit their machine.

The downside was the thick forest between them and the area they needed to get to. It was morning and the fire was still waking from the coolness of the night. Jordan trusted that they had calculated the indices and future fire behaviour correctly. Now was the time to hike the half kilometre through the potentially unavoidable tinderbox that stood between them and the goal.

On the opposite end of the fire, they could see another helicopter with Alpha and Bravo descending under the treeline towards the south helipad. Between them, thousands of smoke vines reached through the canopy on the hillside below searching for the warmth of the sun.

During the crew draft in winter, Jordan had been lucky to pick the crew he wanted, firefighters that sought adrenaline and were willing to try tactics that were sometimes risky but often successful. Gary and Andrew were easy choices. His rookie Quinton, a forestry student from UBC, fit in well with the crew. He was grateful for the three firefighters that wanted to kill this fire as much as he did.

"Let's do this boys," he said as they prepared to hover exit onto the meadow pad. The helicopter came to a hover about a metre above ground. The pilot, unable to see what was under the long grass, could not land without being sure what was there. The pilot's hands and feet remained steady on the controls, keeping the helicopter as motionless as his skills would allow.

The three firefighters in the back seat switched their headsets for their red helmets. They each held up their pointer finger signaling to their leader that they were ready to go. They trained for hover exiting once a month for the last two seasons and now was the time to put their skills to the test.

Jordan pointed to Andrew, placed his fists together, then pulled them apart. Andrew opened the door beside him. The noise and wind from the rotor blades amplified inside the helicopter. Andrew placed his boots onto the skids and slowly got out of his seat. He knelt, grabbed the skid with both hands, and in one controlled motion lowered himself to the ground.

Gary followed suit. Even with his massive frame, his strength helped match the fluidity of Andrew as he left the helicopter. Quinton passed them the hand tools and chainsaw then stepped out onto the skid. Holding his balance, he closed the door then joined his two buddies on the ground.

"Thanks. See you in a bit," Jordan said to the pilot. He exited onto the grassy field and knelt beside his crew giving the thumbs up. All four men left the machine without the slightest movement that the pilot could notice. The helicopter lifted into the sky and disappeared.

As they hiked southeast, the roar of chainsaws echoed through the valley from the Rappel crew working on the northern excursion. The red hats focused on the path in front of them as they tried to keep up with the blue helmet in the lead. They left the safety of the open field and ventured into the thick forest.

They cruised at a pace few could keep up with and soon the forest ended. The river valley presented itself in all its natural splendor. A secondary river flowed to the east emptying into the main river and where these two points met, a triangular rocky outcrop jutted out over a 200-metre drop to the emerald water below. The northern tip of the ridgeline was covered in a charred resemblance of what the forest had been. Less than 50 metres west, the fire had stopped at the retardant line keeping it from the forest they had just walked through.

This was the best place Jordan could find to build a helipad.

"Alright Delta crew. Let's get to work."

19

IC 43

"I observed what was the single most complex initial attack in my 13 years of wildland firefighting."
– Brandon Taylor

Every muscle in my body tensed as the sensation of falling came over me. Warning sirens blasted throughout the machine. My eyes remained closed as I braced for the helicopter to hit the ground.

A jarring sensation rippled through my body as the skids smashed onto the ground. The helicopter bounced back into the air momentarily before falling back to rest. There was no pain. I opened my eyes and everything was as it should. The pilot pressed several buttons silencing the sirens. The roar of the rotor blades took over in the absence of the sirens and the helicopter remained motionless. Seconds passed as everyone on board let the relief sink in. I looked at the pilot for any reassurance that this wasn't his typical landing. He stared through the windshield in silence.

"Everyone okay?" I asked.

"Yeah, we're all good back here," Brando replied.

"Alright, Marco, make this pad as big as you can make it. Brando, go find us some water."

"Copy that."

Everyone except Farley and I exited the machine. With the rotor blades spinning, the rest of the crew unloaded the gear making sure to stay low as they moved into the treeline. They watched our helicopter sluggishly rise from the ground and clear the treetops without incident.

As instructed, Brando and Des dropped their gear, left the helipad, and moved towards the ridgeline to look for the water source we thought we had seen from the air. Marco buckled on the orange chaps preparing for the extensive chainsaw work needed to make the helipad large enough to reduce the crippling anxiety of our pilot.

After the helicopter was airborne, we circled the helipad once more then descended towards the sandbar that bordered the emerald river. We landed easily beside the 1,500 gallon bladder that was left there the night before. Farley and I loaded the bladder into the back of the helicopter, and we lifted off the sandbar back to the south helipad. This time the landing was far smoother, probably because the helicopter was lighter with six less firefighters onboard, or maybe because the pilot knew about the wind change. Either way, I felt a ton safer when we landed. The two of us unloaded the bladder and joined our crew in the treeline. The pilot lifted the helicopter into the sky and began his flight to the Bowen fuel cache to refuel and return with the bucket on.

"How's that water source looking Brando," I said into my radio.

"We found a little bit, but it's mostly just runoff from the retardant drops. Maybe enough to fill one bladder but it's not going to replenish itself,'" replied Brando. It was not the answer I was hoping for but it was still an answer.

The area was rockier than expected. Everything around us was green but just under the moss-covered forest floor was a layer of thick black rock. The forest was vibrant and healthy indicating that the vegetation was receiving adequate moisture. However, the rock layer that made up the ground caused any water to quickly drain into the river below. Underneath the giant black spruce and lodgepole pines that had established their dominance hundreds of years ago, a vast variety of plant life had grown in the shadows of the trees. The carpet of moss and lesser vegetation had claimed the forest floor while shrubs of willow, alder, and red-osier dogwood stood head high, filling the space between the ground and the canopy.

If the fire picked up in that thick forest, it wouldn't be safe for anyone. If an evacuation was required, the helipad would have to be bigger. So Marco, Popp, and Bonesaw began the process of opening the pad. They would have to fell the trees that lined the pad away from the opening. If they fell a tree onto the pad, they would have to buck it up and remove it which would add substantial time to the job. Lucky for us, the trees were relatively straight giving them an advantage when directing their fall.

Farley, Soso, and Dezzie filled up Wajax bags from the retardant runoff. They worked the hottest areas along the fire's edge closest to our helipad.

With the crew set up, Brando and I patrolled the ridgeline where the green forest met the charred aftermath of the fire. Most of the burned ground was smoking with the odd flame breathing for life. With orange flagging tape we marked out two separate paths from the helipad to the black, each around 30-metres long. We had two escape routes established just in case they were needed.[16]

"How did you like that first landing?" I asked with both my eyes and mouth wide open.

"Just about shit my pants. That was the scariest helicopter ride I've ever had," Brando said.

"Yeah man, me too. Let's hope our pilot tightens it up."

Brando and I continued to patrol the fire's edge north along the ridgeline. The fire had stopped for the most part where the ridgeline had been smashed by retardant the day before. This "long-term retardant" that was used could hold the fire back for at least 48 hours and it was doing its job well. The retardant line wasn't 100 percent containing the fire on the ridgeline but was well placed in the areas that would be the hottest. The fire had raced up the slopes and stopped at the retardant, so it was easy for the two of us to follow the fire's edge. Areas where the slopes had given way created funnels for the fire and in those areas the vegetation had been entirely burned. More flames burned along the hillside but were not threatening to burn through the retardant for now.

Brando was one of the most experienced and knowledgeable firefighters that I have had the honour of working with. I was grateful for his input and having him there to bounce around ideas. I also enjoyed his company so for me, he made the perfect patrolling tactician to work with.

After about 15 minutes of bush walking, we could hear a chainsaw up ahead of us. Pushing our way through the treeline, we hiked up a steep section of rock and found Jordan and his Delta crew building a helipad on the farthest northern point. The ridgeline that Brando and I followed travelled north beside the river.

---

16 The black is any area that has been burned. It is a common area to use as safety zones since the fuel has already been removed.

Where we met up with Delta crew, a secondary river ran east into the main river 200 metres below.

"Welcome to Pride Rock," said Jordan. The view was awe-inspiring, and the three of us soaked in the beauty of the view for a few seconds.

The previous day the fire had burned hot at Pride Rock and had left only a charred semblance of the forest that had once surrounded it. Less than 50 metres west, the fire had stopped at the retardant line and this was the best place Jordan could find to build a suitable landing pad. The entire Delta crew were covered head to toe in black ash making it difficult to distinguish who was who. Building a helipad in the black thick dry ash mixed with the sand was giving Delta crew some grief.

Andrew sat on the ground with a disassembled chainsaw, brushing off its components. "Damn saw keeps stalling out. Way too much dust here," he said.

"Take duct tape and cover up some of the filter," I suggested. "It will help keep some of the ash out. I've seen some guys clean dirty filters with gasoline."

"How far is it to the other helipad?" asked Jordan.

I looked down at my GPS and scrolled to the right page. "It's 0.98 kilometres. We can set up a bladder here, but we will need a crew hot-spotting the fire in between us until we can meet up in the middle," I said. That would be a perfect job for Charlie crew when they joined the rest of the Unit on the fire.

"Harold this is Adam," transmitted the voice of one of the lookouts on the other side of the river. "Just wondering if I could be of better use on the fire?"

Adam had made it clear during the morning briefing that he wasn't a fan of his role as lookout. And the firefighters on Pride Rock all laughed after hearing him bring it up again.

"Hey Adam, this is Jordan. Can you wave to me so I can see where you are?" Jordan was peering through his binoculars at the lookout location. He easily spotted Adam across the river valley excitedly waving his arms in the air. "Yeah, I got you. You're right where you are supposed to be. Thanks."

The six of us erupted in laughter once again, especially his older brother Andrew.

Across the river valley to the north, the Rappel crew was finishing their helipad. Jordan had been in contact with the Rappel leader. Both used each other as lookouts, keeping an eye out for fire that could potentially come at them from below.

Another spot fire had occurred overnight 15 metres east of the previous northern excursion and it was now smoking at the base of the treeline. All available helicopter buckets were being used on that side to hold the fire. Satisfied that the line was holding on the north end, Brando and I left for the south pad. About halfway back we could hear Jordan and the Rappel leader talking about how the fire was starting to creep at least 30 metres into the forest and needed more water to hold it.

"Tankers," Brando suggested.

I grabbed the radio from my holster. "Jordan this is Harold. You guys need tankers over there? Might be a while until you get water."

Several seconds past.

"Yeah man. Send it."

At 12:35 I requested tankers and was given an ETA of 13:40 from the air-tanker base in Whitecourt.

The two of us stopped several times to clear the forest floor from hot spots that were burning into the retardant line. With hundreds of spots along the ridgeline, we triaged the areas we felt could wait until more resources could work the line.

After we got back to the south pad, we grabbed Farley and continued south to check out that end of the ridge. We followed the black 30 metres downslope from the south pad and then went south on the downhill side of a rocky gorge that had a 50-metre rock face on the uphill side. The very southern edge of the ridgeline ended at a massive rock face 400 metres from the south pad. This was a perfect natural barrier, and we were confident it could hold the fire. The three of us walked back to the north pad to check for spot fires and to show Farley the fire.

Both helicopters constantly buzzed up and down from the river dropping buckets on the hottest parts of the fire. HFZ, our light helicopter, flew to collect Charlie crew at the Kakwa Lookout. They had driven out gear that we would require but we needed a workforce more than equipment. Charlie's task would be to support Delta as well as to monitor the main ridge or any fire that was trying to break through the retardant. The helicopters were needed to bucket on the fire or move crews so I made the call not to sling in any more gear for the time being.

We arrived at the north pad just in time to have the Bird Dog soar above us. "IC 43 this is Bird Dog 127. Got a group of 802s coming in. Where would you like the drops?"

"Good to see you," I replied. "That excursion on the north side of the river that runs west to east. Can you take care of that for us?"

"I see it. It's going to be a tricky one with these winds but we'll see what we can do."

We lined up on Pride Rock and had the perfect vantage point to watch the airshow. The Bird Dog flew low over the treetops with the siren blaring, warning the Rappel crew below of the incoming drop path as well as showing the skimmers behind where he wanted the drops.

The first 802 appeared from the west and descended close above the treeline. The wings tipped up and down as the pilot fought the winds to keep his plane straight. Seconds before the plane was over the smoke, it released the blood-red retardant and the liquid mushroomed down into the forest below. One by one, the next five skimmers followed and rained down their crimson payloads onto the smoke below them. They did a fantastic job covering the hottest area of the northern spot fire. The skimmers left for a nearby lake they could skim on to fill up their water tanks.

The three of us patrolled back to the south pad trying to contain hot spots that were burning through the line. The flames burned high enough in several hot spots that we needed bucket support to hold the line.

After we felt comfortable leaving the hot spots we continued to the south pad. On the last slope towards the south helipad the trees opened and smoke was rising on the far side of it. "Well that doesn't look right," said Farley. The three of us raced across the uneven ground, weaving between tree stumps and felled trees. A spot fire had breached the uphill side of the retardant south of the helipad and grown into a 75-by-150 metre spot fire.

"All Alpha and Bravo to the south pad. We have a spot fire," I radioed. "Both helicopters, need you at the south helipad with buckets. Drop free and we will clear."

Brando, Farley, and I dropped our bags and pulled the forest floor away from the burning edge, creating a fuel break with our shovels and Pulaski. Bonesaw and Popp arrived minutes later with full Wajax bags and sprayed the hottest of the flames on the opposite side. If the fire burned into the south helipad, we wouldn't have enough water to extinguish the flames and keep the helipad operational.

We swung our hand tools into the vegetation, separating the green from the black. Sparks flew from the tool blades as they met the dark rocks underneath the moss. Soso and Dezzie followed behind, smashing the flames with their hand tools. Large boulders created an uneven landscape adding to the challenge of removing a path of vegetation to form what's called a handline. Building handline is one of the most arduous tasks we can perform and is only as strong as your weakest link. Without words the seven of us worked in unison creating a handline around the spot fire. We triaged the spots where the rocks had naturally contained the flames as we hopped to the burning edges. The two medium helicopters soared over top, showering their bucket loads on the smouldering ground around us.

Several buckets later and with the back-breaking handline underway, we were able to contain the spot fire from growing. Without a better water supply though, Alpha and Bravo would have to stay on the spot fire to keep our helipad safe.

At 15:00, a small blue and white helicopter flew over us. "Who's that?" Farley asked.

"I'm not sure," I said. "Intermediate helicopter over fire 43. This is IC 43. Please identify yourself and your objective," I radioed.

"IC 43 this is assessment machine Whiskey Charlie Tango. We were sent here to check on the status of your fire." The office high-ups wanted to know why I considered the fire out of control and had sent a pair of eyes to see for themselves.

"Yeah we have a good plan in place. If all goes well, we should have a BH by 21:00. Wouldn't mind a flight to check on things from the air," I radioed.

"Copy that. Check us landing."

The smaller helicopter had an easier time landing on our helipad compared to the larger one we had in that morning. I approached the helicopter, ducked down to keep myself clear of the spinning rotor blades, and hopped into the back seat.

"How's it going? Looks like you've got a handle on this," the man in the front seat said over the intercom.

"It's pretty touch and go but we've got this," I said.

The helicopter lifted through the canopy and for the first time in hours I could see the entirety of the fire. Red and blue helmets scurried around the south helipad digging guard or smothering the smoke with their hand tools. Most of the smaller vegetation and about half the trees on the hillside from the south helipad to Pride Rock had been burned off. A distinguishable line of smoke rose along the ridgeline where the fire had burned into the retardant. There was more smoke on the south hillside than I had anticipated. The entirety of the south slope was covered in thick white smoke hiding the dense canopy underneath.

I explained our strategy and why each firefighter was where they were. The south flank was the only part we had not tied in. By using the massive rock face as a natural barrier, we were waiting for the fire to burn into what we hoped would stop that flank.

On cue to test my theory, a tree mid slope torched causing the tree above it to light up. Several trees lit up at once causing black smoke to billow into the air. Forty-metre-tall orange and yellow flames licked the sky. The flames grew into a wall and with each tree gained momentum towards the south rock face. The northwest winds pushed the thick smoke over the rock face obstructing our view of the path of flames. The only hope of stopping the wall of flames was if it burned into the rock face and stopped once it consumed all the fuel in that direction. But if it spotted into the forest on the opposite ridge we would be fucked. In fact, if this fire took a run above the ridgeline there would be no way of stopping it.

20

ALL OR NOTHING

"There are no bad days, just challenging ones."
– Quentin Schmidt

Crimson flames shot high into the air. The wall of flames consumed the trees pushing massive amounts of smoke into the air, obstructing our view of the helipad. Soon after I was airborne, the smouldering hillside had turned into a fiery landscape. The timing of the flight with the provincial wildfire assessor was perfect because we'd see if our strategy of using the south rock face as a natural barrier was going to work.

The fire and smoke grew as the wall of flames devoured the forest. All life that stood between the blaze and the rock face was consumed. Yet the flames disappeared as they met the rock face turning instead into white billowy smoke. The charred trees left standing on the smoking ground were pitch black. I was happy I had not put any crews on that flank and had instead used the natural barriers to our advantage.

We circled above looking for spot fires. No smoke or flames came from the adjacent forested area or in the shale slopes that ran to the river below. The plan had come together. The rock face had held.

"Well, looks like you got things under control. I should get heading back to Grande Prairie before we get smoked in," the assessor said over the intercom, referring to the potential smoke that could come in from the enormous fire northwest of us.

"Any advice before you go?" I asked.

"Nope. You're doing great. Keep up the good work," he replied. After that solid advice, we landed back on the south helipad, and I hiked to the spot fire to check on its containment. The crew completed the last of the handline around the spot fire and the smouldering moss was once again contained. It was 15:15 and the 802s were back from their refill ready for another drop.

"IC 43 this is Bird Dog 127. ETA three minutes and looking for target."

"Bird Dog 127. Looking to cool down the area south of the south pad above the ridgeline. All ground crews are clear."

"Copy that IC. Easy drop."

I have been under drops from the 802s and getting clear of the drop zone was a personal choice. I had been hit by rouge drops before and it was no danger compared to the big tanker loads that can crush the forest below. So I leaned under a large black spruce close to the drop zone. The rest of the firefighters cleared the area to take a water break after busting their asses digging the handline.

The Bird Dog flew low over the canopy, its loud siren yelping overhead indicating that a drop was imminent. The first 802 arrived seconds after. Skimming the treetops, it opened its tank and the water hit the canopy bursting into mist as it fell to the forest floor. The second 802 followed suit and released its load one wingspan from the previous drop. The last four repeated the pattern soaking the entirety of the excursion.

"Bird Dog 127. Great drops," I transmitted.

"IC 43. Thanks. We have to get going before we get smoked in from 44."

"No worries. We'll be able to handle it from here. Thanks for the great work. See you on the next one."

While this was all happening, Charlie crew had completed their task of driving the gear to the nearby lookout. With that completed, they were needed on the line and flew into the south pad where I met them for a briefing. DK took two of his members to work with Delta and left the other two to help Alpha and Bravo crews.

Shortly after Charlie crew arrived, Couse mentioned over the radio that he had seen a small spot fire on the south side of the huge rock face. Hearing this, Brando, Farley, and I bush walked the west side of the rocky gorge and went to see if the spot fire was there. Unfortunately, when the wall of flames had cooked up the far south side, it created firebrands that spotted over the rock face that had stopped the main surge. Light smoke accumulated on the bottom third of the slope putting the south flank in jeopardy. This spot fire had become our new priority. I radioed both pilots and moved their bucket operations to the new spot fire.

"Dammit," I said as I looked at the spot fire below us. The slope was covered in shale but the needle and leaf litter that filled the spaces between the rocks created a problem.

"Man, just say the word and I'll take some of the crew and stop that," Brando said.

"You sure? That looks like it's going to be a bitch to put out."

"If we don't, you know what's going to happen? This will all be for nothing," he reasoned. If the fire caught the right fuel source and found its way to the forest above, all the containment lines would be useless. If this spot wasn't contained it had the potential to race up the hillside and come back around on the firefighters working the south flank.

"Yeah buddy, I'll get some of the crew to meet you here," I said, instructing Farley to stay with Brando. I then radioed Jesse and Couse to meet up with them. "I'm going to go check on the helipad. Both helis are yours. Good luck buddy."

Brando and Farley looked at the slope below and took a breath. It was going to be a struggle to contain this spot fire. With Jesse and Couse arriving minutes later with full Wajax bags, they would have to be strategic with the limited water and rely more on the water that came from the helibuckets.

Brando dug the spade of his shovel into the rocky slope below him to gain some balance as he stepped onto the steep slope. He leaned his weight back as to not topple forward and barrel-roll down the 250 metres of sharp rock into the river. Patches of smoke drifted up from below. Fighting fire from uphill was never a good idea but with only a handful of trees and shrubs spaced out below him, and with only needle and leaf litter in the rocks, he felt safe enough to proceed. Farley, Jesse, and Couse followed Brando off the ridgeline and onto the slope.

When Couse stepped onto the slope, a small rock became dislodged and tumbled down the hillside. It gained momentum and bounced 100 metres until it splashed into the emerald river below.

"Be careful guys. No one be above anyone else, so we don't kill each other," Brando said to the three firefighters that were now his responsibility. The four spread out and began their descent. Couse and Jesse with the added 60 pounds (27 kilograms) of water in their Wajax bags leaned back with each step. Sometimes they even sat to keep from falling forward.

After several minutes scaling down the rocky slope the spot fire presented itself in its entirety. It had grown to 0.3 hectares and was creeping slowly in between the rocks.

"Let's put out the top of this so it doesn't race uphill," Brando instructed. "Farley and I will start flagging out the spot for the buckets."

Jesse and Couse sprayed the top of the spot fire while the other two placed pink flagging tape around its perimeter. With the clear target now in sight for the pilots, they dropped their bucket loads inside the ribboned area. After each drop, the firefighters turned over rocks and spread water onto the smouldering debris.

All bucket support was focused on holding this new spot fire on the shale slope. Brando's muffled voice gave orders to the three men and the two helicopters. His lungs fought for breath between each word. I knew this was going to be my last time fighting a fire on the Unit and I had to make the best of it. I needed to get on the slopes and see what I could do to help.

Popp and I filled up Wajax bags and hustled to the top of the hillside and descended the shale slopes to the four firefighters below. By the time we gingerly made our way 200 metres down the hillside, the fire was contained but smoke crept from between the black rocks. We worked the spot fire for half an hour ensuring it wouldn't spread up the hill and behind our containment line. Legal down was becoming a factor so the six of us climbed back up for extraction back to base.

Breathing heavy and wiping the sweat from our eyes I could see how much the climb had taken out of the firefighters. Brando dropped to one knee and puked. The five of us felt just like he looked. If we hadn't pushed ourselves to our physical limits, the spot fires could have undone the whole Unit's hard work.

We had about an hour until extraction, so Bonesaw and I did one last walk to the north pad. We headed along the ridge to check out the fire and determine next day's priorities.

With Charlie crews hot-spotting, the ridgeline had held beautifully but would still need a ton of work to secure it from spreading.

The north spot fire was contained by the Rappel crew. Delta crew had completed the north helipad and contained the north flank. Alpha and Bravo held the south flank. Echo had been our eyes and kept us from being blindsided from the fire below. With bladders set up on each of the helipads, containing the rest of the fire should fall into place.

After 16 hours of tense, back-breaking work, the combined efforts of multiple tanker pilots, Rappel 2, three helicopter pilots, and the Peace River Unit, successfully held fire 43 to only 31 hectares.

"Dispatch this is IC 43. Check GWF 043 being held at 21:00."

21

THE BEGINNING OF THE END

"For 'Berta."
– Andrew Farley

Brando and I sat on the rock face looking over the river valley. The sky was clear of planes and helicopters. The only sounds came from the wind carving through the trees and mixing with the ambient sound of the flowing river below. The fire was asleep.

A cold front had settled over us and the cooler temperatures that coupled with the higher humidity helped put the finishing touches on fire 43. For three days we held the fire to the slopes. By filling up the bladders on the helipads with the helibuckets, we used thousands of feet of hose to snuff out the fire's edge. Only pockets of heat remained on the hillside that would be left to burn out on their own. The ridgeline was tied in, and the spot fires were extinguished. The Wildfire Twenty patrolled the line, ending any chance the fire had about coming back to life. "I'm going to miss this," I said.

"Yeah buddy, me too. But were not young anymore. Can't be chasing fires forever."

"I don't feel that old." Brando looked at me and grinned. "We'll maybe sometimes." We both laughed.

"At least we picked a good one to go out on, eh? It was amazing what we did here."

Brando was right. It was an incredible feat to hold that fire. GP 43 remains special to me. It showed what we could do as a Unit and the potential for the Unit crew program. The sheer amount of resource allocation, level of strategies and tactics based off the fuel types, fire ecology, reading weather and fire behaviour – it was complex for me at the time. Even though multiple events and decisions were happening at intense intervals, I used my experience and the experiences of those around me to make the best decisions I could in those moments. Being the first one on the line and the last to leave, putting in the hard yards, and not asking anyone to do a task I wasn't willing to do myself – this earned the respect of my crew. Taking the time to work with each crew member in both good and hard times allowed me to learn about their skillsets. Knowing the strengths and weaknesses of each firefighter, I was able to help raise each member of the Unit to their full potential. By building trust with one another, success followed.

Later that year, Jordan entered the data from GWF 043 into a fire growth modelling program to see what would have happened if that wildfire was never contained. With those conditions and fuel types, in all three simulations the fire exploded to thousands of hectares. The amount of resources that would have been required to stop a fire that size would have cost millions.

It was going to be my last fire as the Peace River Unit crew leader, and I made sure to soak in every minute of that deployment. But I knew beforehand that it couldn't last forever. Being on the line is a young person's game. You're gone for months at a time, working crazy hours, and under loads of extreme stress. The body and the mind can only handle so much. I wanted to start a family of my own and I knew that I wouldn't be the type of father and husband I wanted to be if I was gone all the time.

So during that summer I looked for a different job and was successful applying for one of the ten wildfire ranger positions that were created from the same Flat Top review that helped create the Alberta Unit crew program. I didn't want that summer to end. I knew that once I left the Unit, my life would be forever changed and the joy of fighting fires on the ground would become a distant memory. I kept pushing back the start date for my new job until management made me resign as the Peace River Unit crew leader.

My new job was based in Hinton, Alberta, Cavelle's hometown. I spent time with her family there as well as at the Firefighting Provincial Training Centre. It was a beautiful small town located on the east side of the Rocky Mountains and I was looking forward to my new life.

Jordan took over the Unit in my absence and in late summer led the crew for one final campaign fire in the Rocky Mountain District. Like the season before, once the summer was over, each firefighter left the North to go back to their other lives. I followed the Unit's activities for the rest of the summer, adding input when asked, but I knew the Peace River Unit was no longer mine. It was difficult letting go of something I had helped build from the ground up, but was hopeful that the lessons I taught and expectations that I set would help the Unit in future.

Overall that season, the Peace River Unit fought 22 fires in three districts. Our work consisted of issuing fire permits in the spring, prescribed and hazard reductions burns, initial attack of fire flaps, and fighting large fires in the mountains. The crew lived up to my highest expectations. And once again no one was hurt.

For me, the 2014 fire season and the crew I worked with created some of the best memories I have. Not because the wildfires we fought were life changing or that we saved towns from destruction, but because of the friendships that were solidified and the amount of enjoyment I had every day with those 19 other people.

The rookie class of Popp, Julie, Quinton, Bonesaw, and Joel was the best group of new firefighters that I have had the privilege of working with over my wildfire career. The five of them stayed on the Unit for several years, all becoming successful sub-leaders before they left for other paths in life.

Every morning I would wake up and look forward to what the day had in store for us. We believed in the purpose of our job and received satisfaction from the fact that we were accomplishing something that was unique to us all. Every night I would go to bed with a smile on my face and was excited to do it all over the next day. And pushing myself to my physical and mental limits was pleasurable, in a sick sort of way.

I have an emotional time looking back on that point in my life because I don't remember any other time when everyone I was surrounded with believed in the same purpose and what we could accomplish when we held each other to the highest standards. Work harder than you can imagine, care for one another, celebrate the uniqueness of each other, and do it all while having fun. Once again, I was, and I am, proud.

22

REFLECTIONS

"Be the hero of your own story."
– Brandon Taylor

With the 2014 fire season over, and with the success of the Unit crew program, the powers that controlled the Alberta Wildfire entity decided that three more Unit crews would be created for the province. Two of those crews were based in the Whitecourt and Edson districts while a second Unit crew was added to Peace River. With the Rocky Mountain Unit crew established the previous year, plus the Grande Prairie and Salve Lake Units, the province now had a total of seven Unit crews. Even though it was great news that we had impressed our district enough to warrant another Unit crew, it meant that the powerhouse that we had become would be split up between the two Units.

With me leaving for a full-time position in Hinton as a wildfire ranger and Brando transferring to be a man-up supervisor, the two leader positions were given to Jordan and Marco. With 20 new firefighters needed in the district and with DK moving on to chiropractic school, nine of the red hats moved into sub-leader positions and were divided between the two Units. Twenty-three rookies were added to the rosters and the two Unit crews would once again have to start over; however, this time they had a foundation to build from.

Quickly after my transition to wildfire ranger, I began to regret my decision to leave the Unit. Like many wildland firefighters before me, I had given up my career on the fireline for what I thought was love. For Cavelle, I made the compromise to leave the life that I had dedicated myself to over the last 14 years for a desk job where I could be home every night and settle down. After four months of living in my new town, I came home from work one day to find that my then fiancée had moved out. I was heartbroken. I am grateful for the time we spent together but as I learned, nothing lasts forever. With my reason for being in Hinton gone and finding little joy in my new job and surroundings, I toughed it out for another two months, then resigned from the job.

But as a pilot buddy once told me, "Nothing cures all like a good fire season."

And he was right. I moved back to be with my wildfire family in Peace River. With both leader roles taken on the two Unit crews, I joined Brando as a man-up supervisor until a Unit crew leader position would once again be available that I could apply for.

The 2015 fire season ended up being one of the busier fire seasons in the past decade – and the busiest of my career. Large fires burned continuously throughout that summer, so all Unit crews were called to the frontlines non-stop. Both Peace River Unit crews spent their time fighting fires throughout Alberta and the Northwest Territories. Each crew was tested in their own way and faced challenges that came with their individual successes and failures. By the time the summer ended, each Unit had worked over 650 hours on the line while keeping to the high standard that were set before them.

With the need for crews also came the need for experienced firefighters to lead them. I went non-stop from May to September filling in as strike team leader, man-up supervisor, division supervisor, and even as a task force leader when I went to help Idaho with their fires. I gained valuable experience leading crews from five different provinces, as well as Mexicans, South Africans, and Americans. After putting over 850 fireline hours on my legs in five months, I was burned out and relieved when the season was over. I was living with Brando in Edmonton and I remember him telling me at the end of the season that it was clear the fire season had broken me and that he had never seen me so fatigued. "Man, you look worse than when you broke up with Cavelle," was how he put it.

Chronic fatigue is very real. After a busy fire season it takes weeks, sometimes months to recover. Living in a constant fog because you cannot physically or mentally recover fast enough was frustrating, and I wish I knew why that was at the time. Only after the heavy fire seasons of 2017 and 2018 did wildfire operations in this country address chronic fatigue as a medical issue and load management became more actively accepted.

When the 2015 fire season was done, it seemed that every firefighter in Alberta needed a break. During that winter, Jordan finished his degree in forestry at the University of Alberta and got a job in fire-growth modelling for the Provincial Interagency Forest Fire Centre (PIFFC). This left a leader spot open. I quickly showed interest in getting my old position back and after an informal phone call to the Peace River Unit crew coordinator, I was once again the Unit crew leader. Jordan resides in Edmonton, and he remains one of the best friends I could ever have.

The human body can only handle so much and the hard miles that had accumulated on my body caught up with me. Wearing steel-toed boots and at times putting over 30 kilometres a day on the fireline, my right Achilles' developed tendinitis and that was the beginning of the end. The second half of the 2016 fire season was trying to say the least. It hurt to walk and would take hours of stretching everyday just so I could work at a normal pace. I nursed that injury. But because I was a 32-year-old in denial, thinking that the pain would simply go away, my injuries compounded. I knew then that I could no longer be the firefighter that I once was. Fighting fires at that level cannot be sustained and like an athlete or ballerina, a time comes when you must make the conscious choice to slow down before that choice is made for you. I decided that my time leading a Unit crew had once again come to an end and I started training for city fire department jobs so I could stay in the field of emergency response.

I led the Peace River Unit crew for three more seasons until ending my 14-year career with Alberta Wildfire in October 2018. By the time I retired from the Peace River Unit, our crew was holding fires the size of the Fish Lake Fire by ourselves. It was amazing what we were able to accomplish in half a decade. But the best accomplishment over my career was never having anyone that I led get seriously hurt.

When I moved away from the Unit, I was ready to be done leading men and women in the outdoors. And my values in my life had also changed. I wanted a life where I could stay in one place and be happy when the summers got hot, and I could enjoy the heat in a recreational capacity.

While transitioning to my new life on the West Coast, I worked one more fire season as a sub-leader on BC Wildfire Service's Thunderbird Unit crew on Vancouver Island. This finished off my 20-year career as a wildland firefighter. As luck would have it,

Andrew Fulthorpe was on that crew, and I was able to spend my last summer in the forest with someone that I already had a bond with. Andrew now works as a powerline technician in Northern BC. The wildfire family is small and when working in that world it doesn't take long for me to run into someone I know. That summer I also ran into his younger brother, Adam, who had gotten on the Paratack program in BC, firefighters that parachute out of planes into wildfires.

Funny, the Alberta Unit program was moulded after the BC program and my experience on Vancouver Island further validated how far the Peace River Unit had developed. I would put my former crew head to head with any Unit crew in British Columbia. The BC Wildfire Service was a great organization to work for and working for another government has its pros and cons. Not having to wear steel-toed boots while hiking up mountains and through the forest was a huge perk. The Unit crews in BC do not live together on base and after the workday is done, each firefighter goes home to their separate lives. All the sub-leaders were in their 30s and most were married or had children. The closeness that was created from sitting around a campfire at night and living together was not there and I found that I too started treating it more like a job than a lifestyle. That wasn't the case when I led the Peace River Unit.

By the time I retired from wildfire, I had seen my late teens, my 20s, and my early 30s pass by at a rate that quickened each year. I have fought over 315 wildfires and spent more than 8,000 hours on the fireline. I led hundreds of firefighters on the frontlines and hope that I did my absolute best for them. I feel I have given so much of myself to the wildfire community, but in return, I was given so much more. I earned a livelihood and gained experiences that I am thankful for.

When I look back on my time with the Unit, I cannot help but be proud of what was accomplished. What the Unit was created for and what it had become since its beginnings represented an evolution just like that of any other dynamic entity.

The men and women who sacrificed, and continue to sacrifice so much is why the Unit crew program is successful. Without the determination and desire of the leaders, the program could have easily been a failure.

I have seen hundreds of firefighters come and go during my career and with them the reasons for doing their jobs. There are a select few, however, who have done this job because they genuinely love what they do. These individuals who took it upon themselves to rise above the standard expectation and do their best because of their passion – this is how greatness is achieved. Without the help from those people, we could not have accomplished what we did. I am honoured that I got to work with firefighters who put the job and others before themselves. I am privileged to call those people my friends. Jordan, Brando, Farley, Jesse, Gary, and Dezzie, I am forever grateful for your help and for the friendships that we shared.

Most of the firefighters of the Peace River Unit crew, like those before them, moved on with their lives and away from the job that once occupied their summers. Some stayed in wildfire and left for warmer climates while a handful still work on the Peace River Unit protecting the North.

I am happy that my friend Andrew Farley now leads the crew that we built together. From the beginning, I hoped he would one day take over and continue what we had started, and I am glad the Peace River Unit is left in his capable hands.

In 2016, Brando became a wildfire ranger in the Fort McMurray District supervising the Unit and Helitack programs until he reprised his role as a man-up supervisor in 2021. We remain close friends and see each other as much as we can even with the distance between us.

Dezzie married another firefighter and they both completed their diplomas in forestry technology at the Northern Alberta Institute of Technology. They continue to work in wildfire and like the trees around them, grow stronger together.

During spring of the 2017 fire season, Joel got into a fight with a great grey owl and lost the sight in his left eye after the owl's talon found its way into the back of his eye socket. That accident left him unable to stay in Canada because of insurance reasons, and his firefighting career was over. I remember people being terribly upset about the accident as was I, until Jordan put it into perspective. Joel was wild and lived life without fear of consequences, and it was only a matter of time before he was going to lose a limb or even die, so being blinded in one eye was not the worst outcome. The last thing I heard about Joel was that he was working as a high-rise window washer abseiling down skyscrapers back in his hometown of Sydney, Australia.

Jesse left after the 2017 season and moved to British Columbia with his wife. He continues to teach first aid and work as a paramedic in the Interior.

Gary led the other Peace River Unit for a couple of years until he became a police officer. He still misses starry-eyed Matt Dawes.

As for me, I started a new career as a structural firefighter with the Richmond Fire Department on the West Coast in the spring of 2020. This new chapter of my life started during an incredibly stressful time in the world, and I am thankful that I had a full-time job during the pandemic. Some of my buddies were not as lucky as I. It has been fulfilling to help those on a more personal and direct level and be able to go home after each shift.

The trend of hotter and dryer summers means the Unit will be needed more than ever. We are living in a world where record-breaking fire seasons occur at a frequency we have never witnessed. Now with our summers being smoked out for months on end, we are constantly reminded that we are at the mercy of our planet and the forces of nature all around us.

The next time you see a wildfire maybe you will think differently about the crews that are fighting to keep you and your loved ones safe. A blend of young men and women who are trying their best – and at times, figuring it out as they go. It takes a rare type of person with courage to put themselves onto the dangerous frontlines of a wildfire, then sleep in a tent at night, separated from their friends and family. I am grateful that I lived that life for 20 years and to have witnessed wildfire from a perspective that few people have had.

A large part of me still misses the Unit and I often find myself daydreaming about being in the wilderness with my friends. For two decades, it was an all-consuming part of my life and turning that part of my brain off has been difficult. I can't imagine I will ever have that much responsibility or autonomy in my future career and maybe that's not a bad thing. I can now close my eyes at night and not think about the thousands of possible choices that need to be made to raise 20 people to their full potential.

Wildfire still calls to me. I am constantly torn between living in a city and moving into the wilderness. Perhaps one day I will find my way back to what I love.

After all, nothing cures all like a good fire season.

EPILOGUE

Good timber does not grow with ease:
The stronger wind, the stronger trees;
The further Sky, the greater length;
The more the storm, the more the strength.
By sun and cold, by rain and snow,
In trees and men good timbers grow.
– Douglas Malloch

On the morning of November 10, 2021, I boarded a plane in Vancouver destined for Edmonton. It had been over two years since I had been on a flight because of the pandemic but it was important for me to make the trip. Exiting the Edmonton airport terminal, I zipped up my coat and threw on a pair of gloves. The weather was reasonably warm for that time of year, but I was cold. Three years of living on the West Coast had acclimatized me to the mild winters and I momentarily wondered how I spent so many years up here in Northern Canada. I waited in the pickup area and after a few minutes, a silver Volvo pulled up. Through the windshield shone a huge grin surrounded by a thick auburn beard. The tall man got out of the car and hugged me.

"Man, it's good to see you buddy," said Jordan. Since the beginning of the pandemic, I had only seen my friend once and it had been over a year since we last hung out.

"Yeah, me too. Too bad that it had to be for this reason, but it's good to see you," I replied as I got into his car. For the next hour and a half, we drove south to our destination in Red Deer. We talked frequently over the phone since my move to the West Coast, but having an in-person conversation was so much more enjoyable than catching up long distance.

We arrived at the Catholic Sacred Heart Church and parked on the street. It was the tail end of autumn and the leaves had disappeared from the trees. Snow had yet to fall but the air was brisk, hinting that winter could come at any time.

A tall slender man dressed in a dark suit got out of the white truck parked in front of us.

"Hey, it's Brando," I said to Jordan. I walked over to my friend and wrapped my arms around him. I had also not seen him in over a year and to be around him was lifting. His wavy hair now consisted mainly of greys and the hard years in the bush seemed to be taking a toll on his body.

The three of us walked to the church parking lot. Dozens of unfamiliar people in suits or black dresses filtered into the church. A handsome man in a long grey coat stood out amongst the crowd. It was Gary. I had seen him a month prior when he came for a visit in Vancouver. It was good to see my friend.

"Hey Gary. Let me know how it goes in there and we'll see you after," I said. Because of the Alberta public health measures only 50 people were allowed in the church at the same time so the three of us would have to wait outside.

Dozens of vehicles were parked on the far side of the parking lot. A muscular blond Viking and a tiny woman with hair as thick as a horse's mane stood beside a paramedic truck. It was Jesse and Dezzie. They would also have to wait outside with us.

Over the next several minutes, more and more people that I had worked with on the Peace River Unit arrived, some that I hadn't seen in years. Each person embraced one another and said how great it was to see each other.

The years since I had seen my crew had aged them all in their own ways. Some seemed relatively unchanged while most represented middle-aged versions of themselves. Most still sported the moustaches that they had cultivated during their wildfire career.

After an hour of catching up, we followed the crowd that exited the church to the Mount Calvary Catholic Cemetery. We gathered as six men, three of them brothers, went to the back of the hearse and lifted a pale blue coffin out of the back. They six men carried the coffin to the hole that was dug and placed it on the dark green straps that covered the opening.

Two clergy men read passages from a bible then sang a hymn. When they finished, we watched as the coffin that held our brother James Williams was lowered into the earth.

My eyes welled up and with each blink tears ran freely down my cheeks. Across from me, Farley and Popp stood beside each other, emotion showing openly on their faces. Marco, DK, and 28 other past and present firefighters stood together as we collectively said our silent farewells to the man that was a good friend to us all.

The week before, James, at the age of 32, was found dead in his bedroom. His family kept the details of his passing to themselves but either way – our brother was gone.

James was special to me, and I am grateful for the memories that we shared. His love for adventure and readiness to help those in need stand out in my mind. I wrote this book before he died so my stories are how I remembered him in life. He was a great man and firefighter. I will miss James. Even during the toughest of times, he was positive and lifted the morale on the crew. His passing showed how the family that we became would be a part of us for the rest of our lives.

The firefighters and I stayed the night in Red Deer to celebrate the life of James. We shared our favourite memories of our friend and the joy he brought to the crew.

The last year and a half were difficult for everyone in their own way and for most of us, this was the first gathering outside of local health regions that we had experienced in a long time. We were there not only for closure but to support one another.

It had been over eight years since the first of us fought fires together on the Peace River Unit. After all that time and the distance between us, we remain close. I will always be proud to have been a part of the Wildfire Twenty.

Printed in the USA
CPSIA information can be obtained
at www.ICGtesting.com
JSHW021939060724
65864JS00001B/31